# SpringerBriefs in Molecular Science

For further volumes:
http://www.springer.com/series/8898

Shu Wang · Fengting Lv

# Functionalized Conjugated Polyelectrolytes

## Design and Biomedical Applications

 Springer

Shu Wang
Fengting Lv
Organic Solids Laboratory
Institute of Chemistry
Chinese Academy of Sciences
Beijing
People's Republic of China

ISSN 2191-5407          ISSN 2191-5415   (electronic)
ISBN 978-3-642-40539-6   ISBN 978-3-642-40540-2   (eBook)
DOI 10.1007/978-3-642-40540-2
Springer Heidelberg New York Dordrecht London

Library of Congress Control Number: 2013947378

Printed on acid-free paper

Springer is part of Springer Science+Business Media (www.springer.com)

# Preface

The Nobel Prize for Chemistry in 2000 was awarded to Alan J. Heeger, Alan G. MacDiarmid, and Hideki Shirakawa *"for the discovery and development of conductive polymers."* Along with the development of conductive polymers, various conjugated polymers have been synthesized during the past three decades, which exhibit excellent electrical properties. As an important soluble form of conjugated polymers, conjugated polyelectrolytes (CPEs) bearing ionic pendant groups have emerged as one of the most important classes of transduction materials, which combine the properties of polyelectrolytes, coordinating electrostatic forces with oppositely charged analytes, with the attractive optical and electronic properties of conjugated polymers. The delocalized electronic structure of CPEs impart efficient coupling between optoelectronic segments, and excitons can be efficiently transferred to lower electron/energy acceptor sites over long distances to super-quench the fluorescence of conjugated polyelectrolytes or amplify the signals of acceptors. Properties of CPEs can be tuned by rational design and synthesis at the level of the conjugated repeat unit and pendant groups which endow CPEs with new applications. The integration of CPEs with biomedical research has already shown profound effects on the diagnosis for the early detection of disease-related biomarkers, cell imaging, and disease treatment, which leads to a multi-disciplinary scientific field in the context of chemistry, polymer science, materials, biology, and medicine with an integrated perspective into both basic research and application issues.

The aim of this Brief is to summarize and highlight the rapid expanding area of applied CPEs to biomedical applications. This Brief collects original contributions and review papers on versatile biomedical platform provided by CPEs, starting with a brief introduction to CPEs and various methods developed for the preparation of advanced CPEs, followed by their applications in biological sensing, disease diagnosis, cell imaging, drug/gene delivery, and disease treatment.

The authors are indebted to all the co-workers in our own laboratory for their talents and hard work, and also numerous co-workers from other laboratories around the world with whom we have worked on different aspects of biomedical applications of CPEs, which contribute a lot to the development of new materials aimed at improving health care and advancing medical research.

Beijing, China                                                                                           Shu Wang
                                                                                                            Fengting Lv

# Contents

# Chapter 1
# Introduction

**Abstract** Conjugated polyelectrolytes (CPEs) are generally composed of two important components, which are π-conjugated backbones and water-soluble ionic side chains. CPEs combine the properties of polyelectrolytes, which coordinate electrostatic forces with oppositely charged analytes, with the attractive optical and electronic properties of conjugated polymers. Specific biomedical application of functionalized CPEs can be achieved through delicate molecular design at the level of the conjugated repeat unit and pendant groups. In this chapter, the developed methods for the preparation of functionalized CPEs are presented, and their sensing applications are briefly demonstrated according to the different types of functionalized CPEs; furthermore, the detection mechanisms are also illustrated.

**Keywords** Conjugated polyelectrolytes • Molecular design • Signal amplification • Polydiacetylene • Poly(thiophene) • Poly(p-phenylenevinylene) • Poly(p-phenyleneethynylene) • Poly(fluorene-co-phenylene)

## 1.1 General Introduction to Conjugated Polyelectrolytes

The Nobel Prize for Chemistry in 2000 was awarded to Alan J. Heeger, Alan G. MacDiarmid and Hideki Shirakawa "for the discovery and development of conductive polymers." Conductive polymers have both electrical and optical properties similar to those of metals and inorganic semiconductors, and meanwhile they exhibit the attractive properties associated with conventional polymers, such as ease of synthesis and flexibility in processing [1]. Along with the development of conductive polymers, various conjugated polymers have been synthesized during the past three decades, which show excellent electrical

S. Wang and F. Lv, *Functionalized Conjugated Polyelectrolytes*, SpringerBriefs in Molecular Science, DOI: 10.1007/978-3-642-40540-2_1, © The Author(s) 2013

properties. Conjugated polymers are characterized by a backbone with a delo-calized electronic structure, which imparts certain conjugated polymer with interesting optical properties. These unique optoelectronic properties allow con-jugated polymers to be used for a large number of applications, including light-emitting diodes [2, 3], photovoltaic cells [4, 5], field effect transistors [6, 7], and chemical and biological sensors [8–13]. The chemical structures of conju-gated polymers offer several advantages as the responsive basis for chemical and biological detection schemes based on optical methods. The conjugated units in close proximity of the polymer backbone allows for efficient electronic coupling and therefore fast intra- and interchain energy transfer, which indicated that an environmental change at a single site can affect the properties of the collective system, producing large signal amplification [14]. Most unfunctionalized con-jugated polymers are intractable (i.e., insoluble, infusible) because of the back-bone rigidity intrinsically associated with the delocalized conjugated structure. Generally speaking, both chemical and physical properties of a polymer mate-rial may change with substitution. Consequently, soluble forms of various con-jugated polymers have been prepared by grafting suitable side groups and/ or side chains along their conjugated backbones. Conjugated polyelectrolytes (CPEs) are conjugated polymers with water-soluble ionic sides, which inherit the attractive optical and electronic properties of conjugated polymers and water solubility that is essential for interaction with biological substrates such as DNA, proteins, and microorganisms for further biomedical applications [15]. According to the charge sign, water-soluble CPEs can be simply divided into two categories, cationic CPEs and anionic CPEs. Cationic groups of CPEs are usually quaternary ammonium, while anionic groups of CPEs are carboxylate, phosphonate, or sulfonate. On the basis of the backbone structures, CPEs can be classified into several types, namely, polydiacetylene (PDA), poly(thiophene) (PT), poly($p$-phenylenevinylene) (PPV), poly($p$-phenyleneethynylene) (PPE), and poly(fluorene-co-phenylene) (PFP). The chemical structures of some typical CPEs are summarized in Fig. 1.1. Cationic CPEs have been proven to be very useful for DNA detection based on electrostatic interaction with the negatively charged phosphate backbone, which provide solid foundation for diagnosis of disease-related DNA biomarkers. The combination of strong light-harvesting ability and the multivalent interactions toward multicharged biological targets endows CPEs with remarkable advantages in nonspecific in vitro (e.g., cell level) and ex/in vivo (e.g., animal level) imaging. By functionalizing CPEs with specific recognition ligands, targeting identification and imaging have also been successfully achieved. As for delivery systems, CPEs can offer extremely stable hydrophobic backbones to encapsulate and deliver cargo, and be able to nonin-vasively and real-time monitor cargo loading, delivery, and release by the self-luminous property of CPEs. In addition, CPEs can sensitize oxygen to generate singlet oxygen and other reactive oxygen species (ROS), therefore the potential of CPEs in photodynamic therapy (PDT) has been desirably exploited, which demonstrate remarkable biocidal or tumoridal activity.

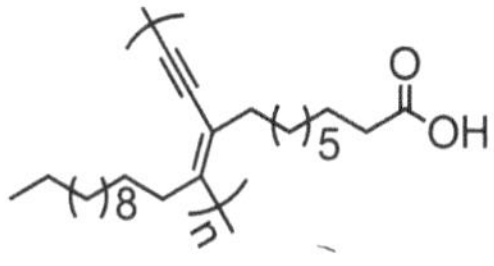

**Fig. 1.1**   The molecular structures of typical water-soluble CPEs 1–7

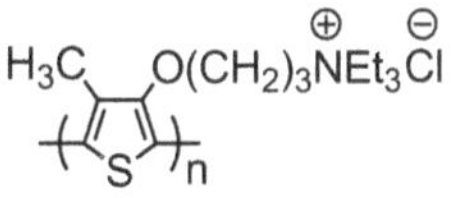

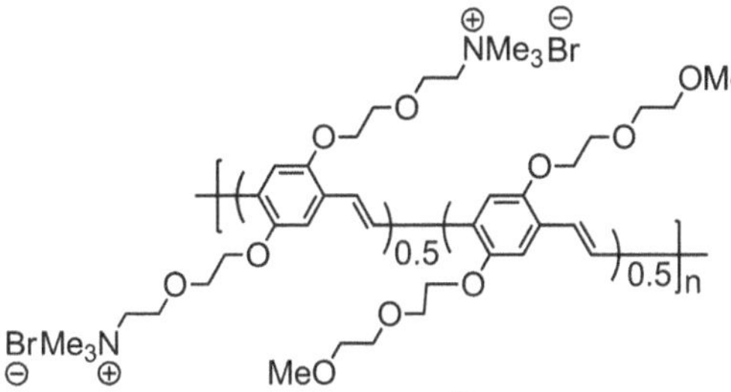

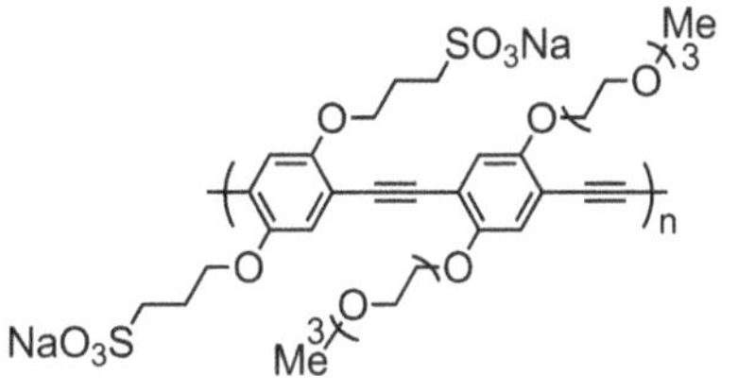

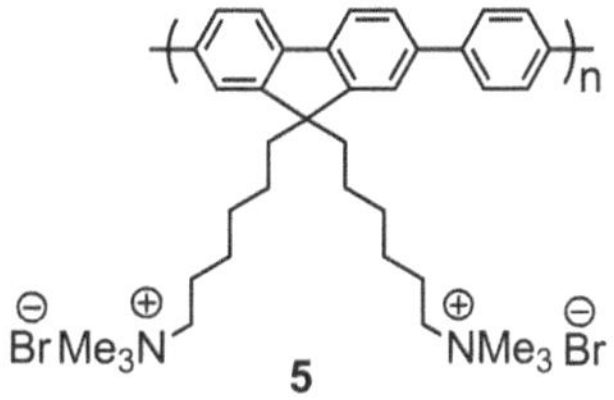

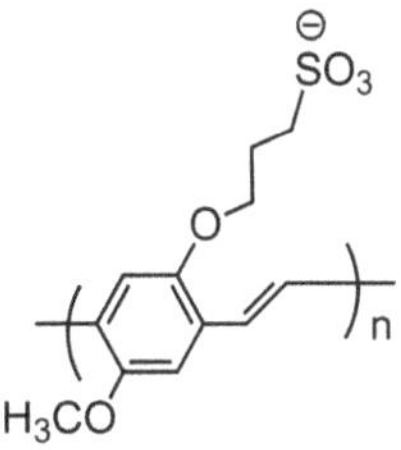

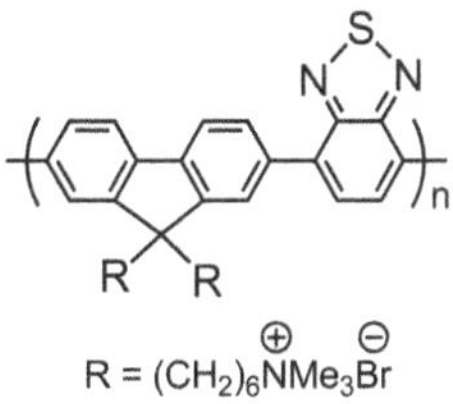

## 1.2  Design and Properties of Functionalized Conjugated Polyelectrolytes

The classical synthesis routes of various water soluble CPEs were exhaustively introduced in the literatures. The commonly employed reactions for the synthesis of CPEs are palladium-catalyzed coupling reactions (Suzuki coupling reaction for PFP [16], Heck coupling reaction for PPV [17], and Sonogashira coupling reaction for PPE [18]), Wessling reaction for PPV [19], topopolymerization reaction for PDA [20], and FeCl$_3$ oxidative polymerization for PT [21], respectively. The emphasis of this Brief is on the molecular design and biomedical applications of functionalized CPEs, therefore references are provided to sources that contain more detail on published procedures for synthesis of CPEs. Structurally, CPEs are generally composed of two important components: (i) π-conjugated backbones, which determine the main optical properties of CPEs, such as absorption and emission spectra, light-harvesting ability, and quantum yield (QY); (ii) side-chains with charged groups. Conjugated polymers functionalized with polyalkyl ether chains, crown ether, and aza crown ether moieties have been the most thoroughly studied covalently modified systems [8, 9]. Inspired by these studies, CPEs covalently modified with a variety of recognition elements (e.g., sugar, small-molecule ligand, and antibody) were developed. For example, Bunz and co-workers functionalized PPEs with α-mannose to achieve the selective binding with concanavalin A (Con A) and bacteria [22, 23]. The synthesis of α-mannose PPEs was displayed in Fig. 1.2, in the last step the pre-PPE was desilylated and a copper cata-lyzed 1, 3-dipolar cycloaddition was performed to give α-mannose PPE 8 in a high yield [24, 25]. Polymers can be designed with folic acid grafted onto the polymer side chain that can specifically kill folate receptor over-expressed cells, thereby pro-viding an important demonstration of anticancer specificity through molecular design [26]. The coupling reaction of folic acid with PPE was carried out in the presence of the carbodiimide EDC (1-ethyl-3-(3′-(dimethylamino)propyl) carbodiimide) in

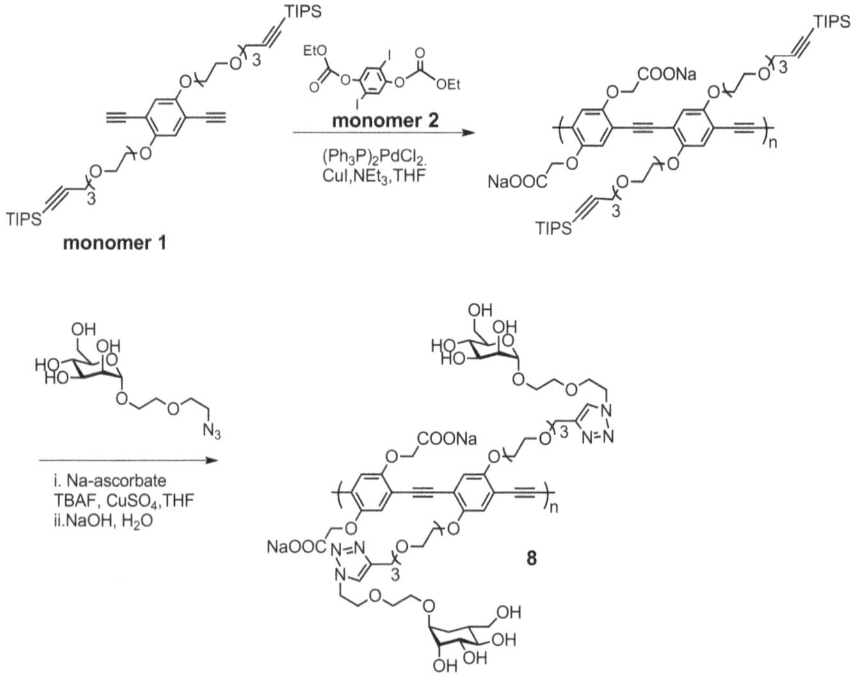

**Fig. 1.2** Synthesis of the α-mannose PPE 8 via Pd-catalyzed coupling of monomer 1 to mono-mer 2 followed by dipolar cycloaddition

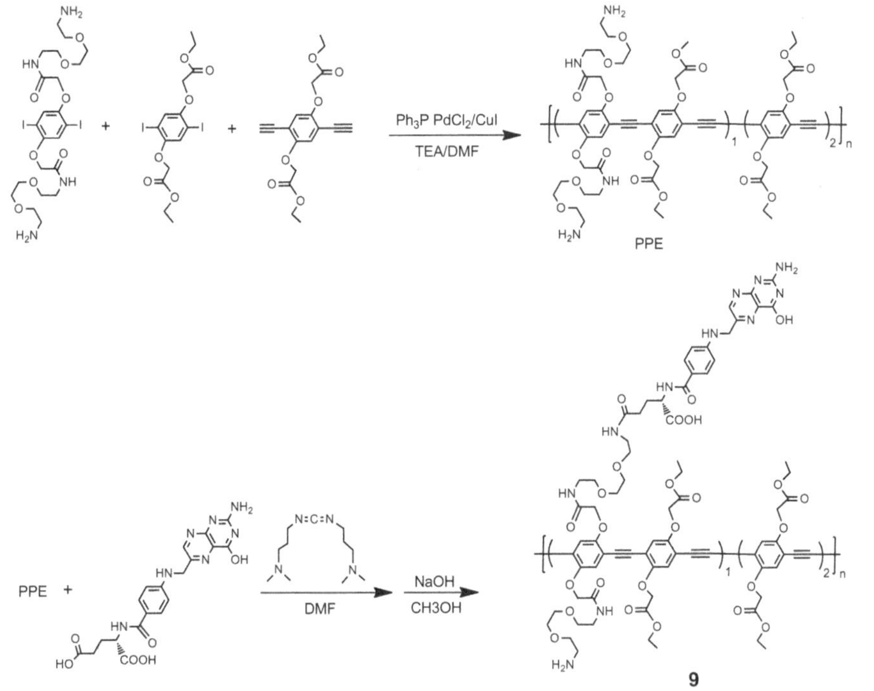

**Fig. 1.3** Synthesis of the folic acid functionalized PPE 9

anhydrous DMF. PPE 9 was obtained after hydrolysis of the ester groups by NaOH in methanol (Fig. 1.3). By using identical coupling reaction, the hybrid CPE-antibody conjugates were prepared by means of direct bioconjugation between an antibody (CD 3 or CD 20) and a conjugated PPE derivative (PPE 10 or PPE 11, chemical structures shown in Fig. 1.4) having blue or red fluorescent emission for fast, convenient, and highly sensitive live cell imaging [27].

By virtue of electrostatic interaction between CPEs and oppositely charged ligands, the self-assembly of CPE 12 and peptide were prepared and realized receptor-targeted cellular imaging [28]. Analogously, CPE 13 (chemical structure shown in Fig. 1.4) containing fluorene and boron-dipyrromethene repeat units in the backbones that exhibits red emission was synthesized. Cationic CPE 13 forms uniform nanoparticles with negatively charged disodium salt 3, 3′-dithiodipropionic acid (SDPA) in aqueous solution through electrostatic interactions, which can sensitize the oxygen molecule to readily produce ROS for rapidly killing neighboring bacteria and cancer cells [29].

Copolymerization of CPEs with various segments provides an alternative way to obtain new function of CPEs. The combination of optoelectronic properties characteristic of conjugated structures and other segments into a single copolymer chain should, in principle, lead to a material with properties characteristic of both the constituent components. A specific example is PFPB, where

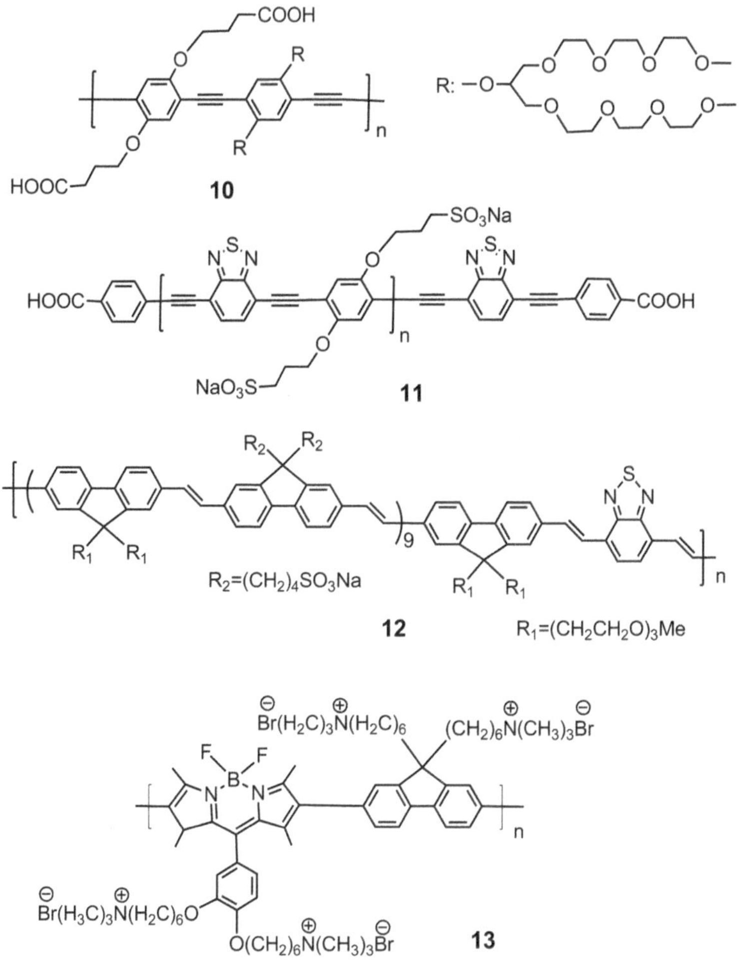

**Fig. 1.4** Chemical structures of CPEs 10–13

the benzothiadiazole (BT) fragment emits at lower energies relative to the flu-
orene-*co*-phenylene segments. It is important to note that Förster resonance
energy transfer (FRET) along the backbone in individual chains is not effective,
presumably because of constraints from a nonoptimal orientation factor, although
this mechanistic point remains unclear. At high PFPB concentrations, interchain
FRET is possible and the emission color changes to that characteristic of the BT
units. Mixing these polymers with specific targets, for example, an oppositely
charged polyelectrolyte causes an increase in the local concentration of the CPE
and triggers interchain FRET. Thus, the emission from the CPE changes from
that of the majority of the backbone to that characteristic of the internal acceptor
units. Specifically, under aggregation conditions, the BT-containing cationic PFP
derivative 14 synthesized by Liu and Bazan [30] and anionic PFP derivative 15

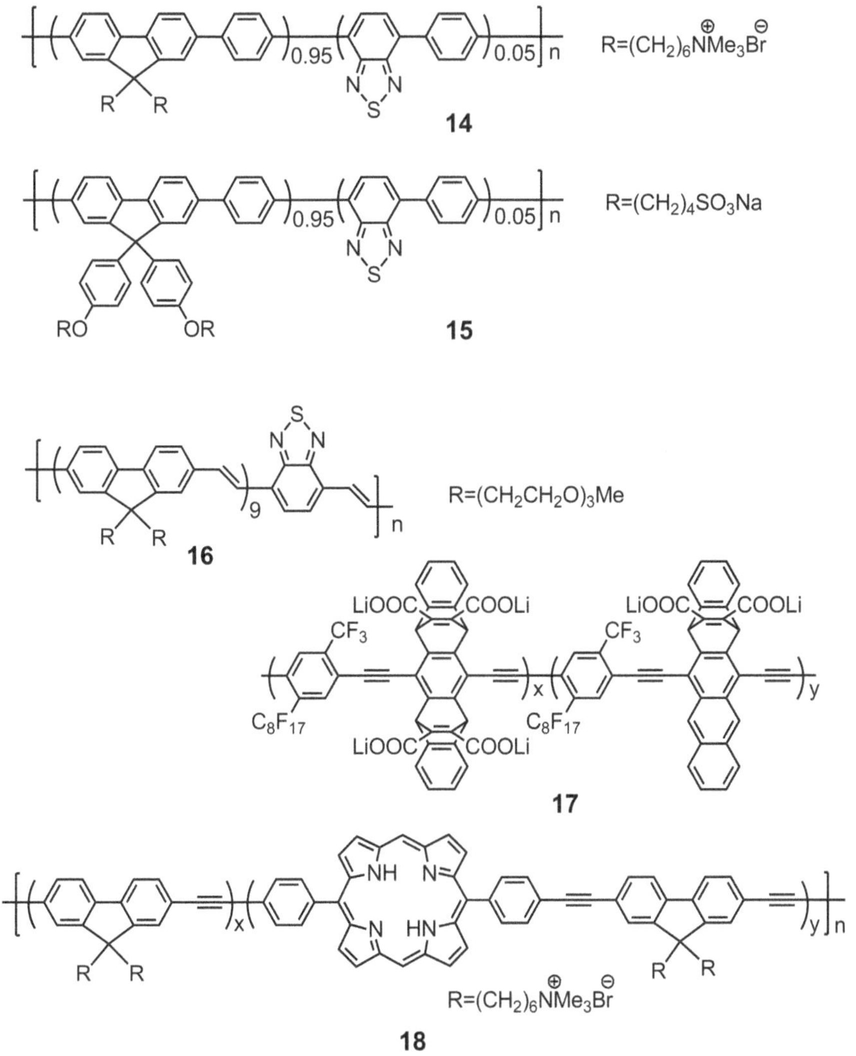

**Fig. 1.5** Chemical structures of CPEs 14–18

subsequently synthesized by Wang and co-workers exhibited a distinct emission color change from blue to green [31]; the BT containing polyfluorenyldivinylenes (PFVs) 12 and 16 synthesized by Liu and co-workers displayed an emission color change from green to red [28, 32]. By incorporating green emitting exciton-trapped anthryl units into anionic PPEs, Satrijo and Swager reported another type of CPE 17 that exhibited aggregation-enhanced energy transfer [33]. Upon ana-lyte induced aggregation, the polymer displayed a remarkable blue to green fluo-rescence color change. Taking advantage of this unique property, the sensing of multicationic amines was realized. In 2008, Liu and co-workers designed a series of cationic porphyrin-containing conjugated PFEs 18 (chemical structure shown in Fig. 1.5) [34], the incomplete intramolecular energy transfer from fluoreneethy-nylene to porphyrin units resulted in dual emission of blue and red.

Another interesting area closely related to the block and/or graft conjugated copolymers is the synthesis of hyperbranched/dendritic supramolecules with con-jugated moieties. Dendritic macromolecules are molecular species with a large number of hyperbranched chains of precise length and constitution surround-ing a central core [35]. Dendritic polymers differ from conventional linear poly-mers in that the former possess much more free volume and terminal end groups, and hence become more soluble than the latter. Liu and co-workers synthesized a water-soluble hyperbranched CPE (HCPE 19) for live cell imaging [36]. The preparation of HCPE was conducted through the combination of alkyne polycy-clotrimerization and alkyne-azide click reaction, and the procedures were given in Fig. 1.6. In addition, the state-of-the-art synthetic methods used provide the

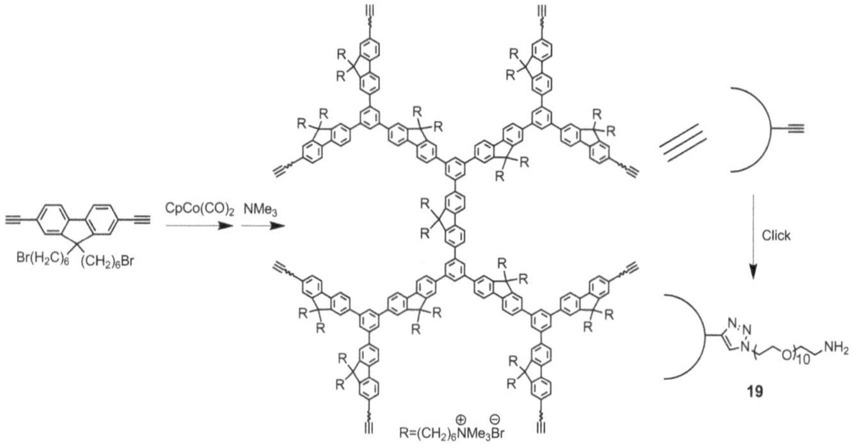

**Fig. 1.6** Synthesis of the water-soluble hyperbranched CPE 19

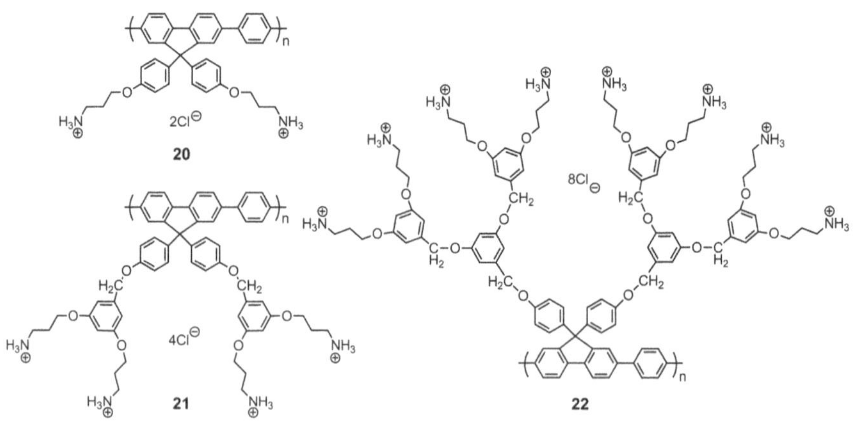

**Fig. 1.7** Chemical structures of CPEs 20–22

feasibility and flexibility to modify both core and shell components of HCPE for specific biological applications.

Water-soluble dendritic poly(fluorene) with positively charged amine groups on the exterior (20–22, chemical structures shown in Fig. 1.7) were synthesized by Wang and co-workers [37]. By virtue of the higher cationic charge density, these novel cationic CPEs is less aggregated in aqueous solution and exhibits an improvement of fluorescent DNA assays signal, whereas higher generation is not easily available due to the steric hindrance of the side chain and the lower yield. In addition, dendritic poly(fluorene) was successfully utilized as gene carriers, which would be discussed in detail below.

## 1.3  Applications of Conjugated Polyelectrolytes

The research efforts worldwide are developing new materials aimed at improving health care and advancing medical research. Interest is booming in biomedical applications for use outside the body, such as diagnostic sensors and "lab on-a-chip" techniques, which are suitable for analyzing blood and other samples. The past decades have witnessed the evolution of molecular design to address the ever-changing and challenging needs of molecular imaging, drug delivery, and cancer therapy research.

CPEs have established themselves as useful optical platforms to sensitively detect chemical and biological molecules. CPEs are characterized by a delocalized electronic structure that exhibits efficient coupling between optoelectronic segments. Excitons can be efficiently transferred to lower electron/energy acceptor sites over long distances to super-quench the fluorescence of CPEs or amplify the signals of acceptors. Among the CPEs, PDA and PT display the conformation-sensitive property to initiate discernible signal changes by naked eyes [12, 38]. PDA is extremely sensitive to external stimuli, such as temperature, pH, mechanical force, organic solvent, and specific ligand—receptor interactions occurring at the surface of PDA matrix [38]. The obtained PDA is blue in color (long absorption wavelength) and nonfluorescent, that is, nonfluorescent "blue state." When PDA suffers from environmental perturbations, both blue-to-red colorimetric and nonfluorescent-to-fluorescent transitions can take place, producing the fluorescent "red-state" PDA (Fig. 1.8). Such stimulation-responsive color change and fluorogenic properties make PDA a very promising sensing material. Similarly, PTs are known to exhibit interesting chromic features (change of color induced by a conformational change of the conjugated backbone) in presence of different stimuli [39]. When a random-coil nonplanar conformation is favorable, PTs exhibit short-wavelength absorption and emission (yellow state, Fig. 1.8). In the presence of exogenous binding targets, PTs are forced to adopt a more planar aggregated conformation via electrostatic and hydrophobic interactions, leading to the red-shifted absorption and emission spectra as a result of the extended conjugation length (pink-red state). Because of the conformation transition from isolated species to

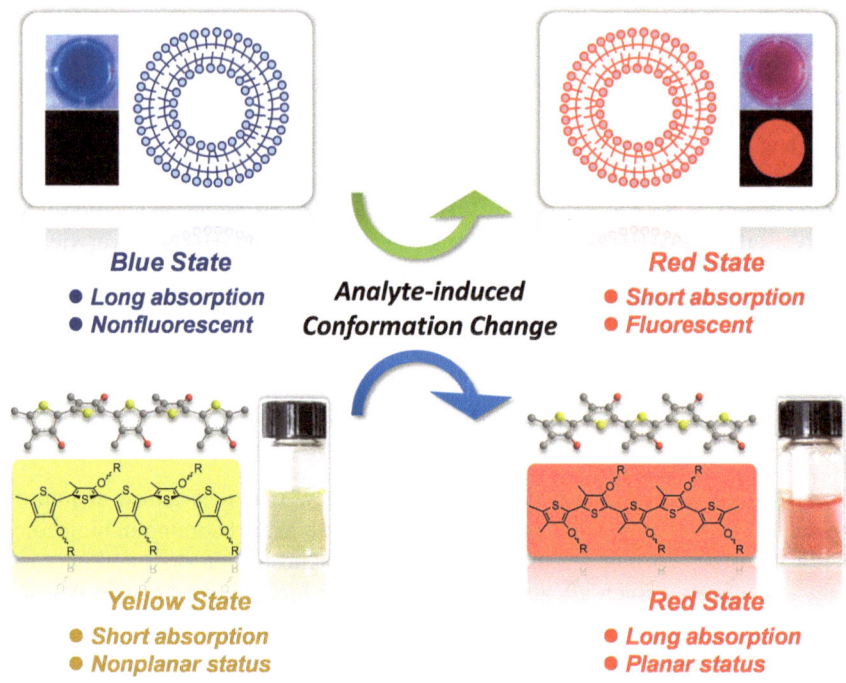

**Fig. 1.8**   Analyte-induced conformation changes of PDA and PT

aggregates, the fluorescence intensity is concomitantly decreased. Therefore, conformation changes could be determined by monitoring the changes of absorption and emission spectra of PTs or direct observation by naked eyes.

The Leclerc group pioneered DNA detection with PT in homogenous solution, which relies upon conformational changes of the PT backbone upon ssDNA hybridization with complementary ssDNA in 2002 [21]. PT 23 (chemical structure shown in Fig. 1.9) can easily transduce oligonucleotide hybridization with a specific 20-mer capture probe into a clear optical (colorimetric or fluorometric) output. Building on this initial concept, in 2005 they reported the detection of DNA targets in aqueous solutions at the zeptomolar range using a physical amplification method described as fluorescence chain reaction (FCR), which is a combination of electrostatic interactions, chromism, and a FRET mechanism, leading to novel fluorescence signal amplification detection [40]. The sensing mechanism was also extended to protein detection [41].

Various diagnostic methods toward bacteria and cancer cells have been developed utilizing the specific conformation changes of PDA and PT, which will be demonstrated in the following section concerning biomedical application of CPEs.

An important characteristic of CPE materials is the efficient quenching of the CPE fluorescence that can be achieved by quencher-labeled target through either electron transfer or energy transfer. This amplified quenching ("superquenching") can be attributed to a combination of Coulombic and hydrophobic interactions

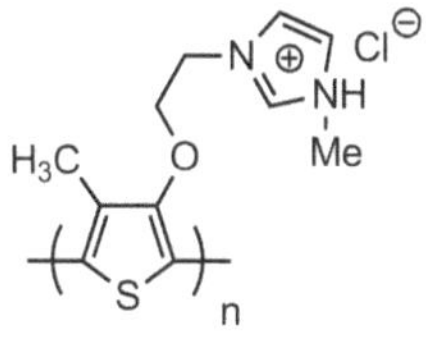

**Fig. 1.9** Chemical structures of CPEs 23 and 24

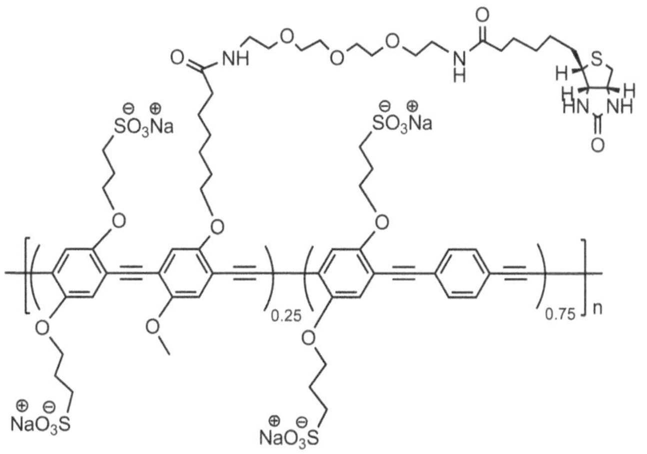

in aqueous media and to rapid migration of excitons through the polymer following photoexcitation [10]. Trace detection of analytes including ions, biomacromolecules, and explosives have been successfully accomplished by utilizing this amplification mechanism [8]. PPE derivatives as one of the representative CPEs were employed to develop DNA sensors by this superquenching mechanism [42–44]. Whitten and co-workers developed DNA sensors employing biotinylated PPE (PPE-B, 24, chemical structure shown in Fig. 1.9) for a target single strand 20-base sequence of DNA coding for the anthrax lethal factor (ALF) [42]. Addition of the target strand would not attenuate the fluorescence of PPE-B, whereas addition of a quencher-labeled target strand (DNA-QTL) would reduce the fluorescence of the sensor. Schanze and co-workers have examined fluorescence quenching of the anionic conjugated polyelectrolyte with a series of cationic cyanine dyes to provide insight into the mechanism of amplified quenching [45]. In addition, the light-activated biocidal activity of cationic PPE derivatives and oligo(phenylene ethyneylene)s (OPEs) were further exploited utilizing the electrostatic interaction between cationic CPEs and microorganisms, which will be discussed in anti-microorganism application.

The quencher can be designed as an energy acceptor to allow for efficient fluorescence resonance energy transfer from the CPEs to the acceptor. The unique characteristic of CPEs is their ability to amplify signals of dyes labeled on analytes, therefore lower concentration of analytes can be easily detected, in comparison to directly excite the small molecular dyes. DNA hybridization assays employed FRET of CPEs have been well developed and extensively investigated in Bazan group [46–50]. The initial demonstration of specific DNA sequence detection involved a fluorescein (Fl)-tagged peptide nucleic acid (PNA) probe (PNA-Fl), and poly(9, 9-bis(6'-N,N,N-trimethylammonium)-hexyl)- fluorenephenylene) containing iodide counteranions (PFP-NI), as illustrated in Fig. 1.10 [46]. Hybridization of the neutral

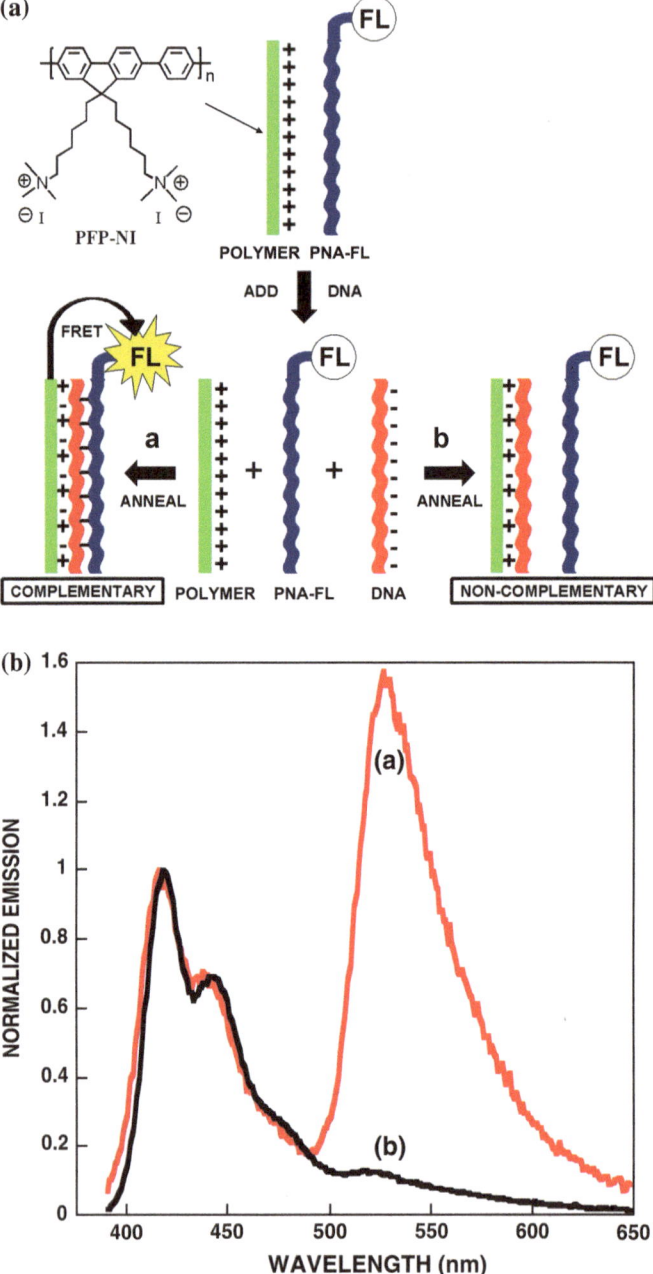

**Fig. 1.10** Schematic representation for using PFP-NI with a specific PNA-Fl optical reporter probe to detect a complementary ssDNA sequence (*up*). Emission spectra of PNA-Fl in the presence of complementary (**a**) and noncomplementary (**b**) DNA by excitation of PFP-NI. The spectra are normalized with respect to the emission of PFP-NI (*down*) (Reprinted with permission from Ref. [46]. Copyright (2002) National Academy of Sciences, USA)

PNA-Fl to the complementary ss-DNA results in the formation of a PNA-Fl/ssDNA duplex, which exhibits net negative charges. Electrostatic attraction between PFP-NI and the PNA-Fl/ssDNA duplex brings fluorescein sufficiently close to PFP-NI to favor FRET (situation a in 1. 10). One, thereby, observes a 25-fold increase in fluorescein emission relative to direct excitation of fluorescein at its absorption maximum. When the target is not complementary to the PNA probe, hybridization does not occur, PFP-NI to fluorescein distance is large, and almost no fluorescein emission is observed (situation b). Figure 1.10 shows emission spectra of PNA-Fl in the presence of complementary (curve a) and noncomplementary (curve b) DNA by excitation of PFP-NI. A wide range of targets especially diseases related biomarkers have been successfully detected via this approach, which take advantage of FRET as a means to modulate fluorescent responses.

Chemical nose/tongue-based sensor array approaches that exploit differential receptor-analyte binding interactions provide an alternative to lock-and-key principles that use specific recognition processes [51]. The major advantage of these sensor array approaches is that they do not require prior knowledge of unique molecular signatures. A common strategy for these approaches involves the construction of detecting elements arrays and collecting the combined response (FRET or quenching signals) of the library in which each component responds differently in the presence of a specific target. Generally, for chemical nose/tongue strategy, mathematical analysis methods, such as linear discriminant analysis and principle component analysis, are necessary to transform the response patterns

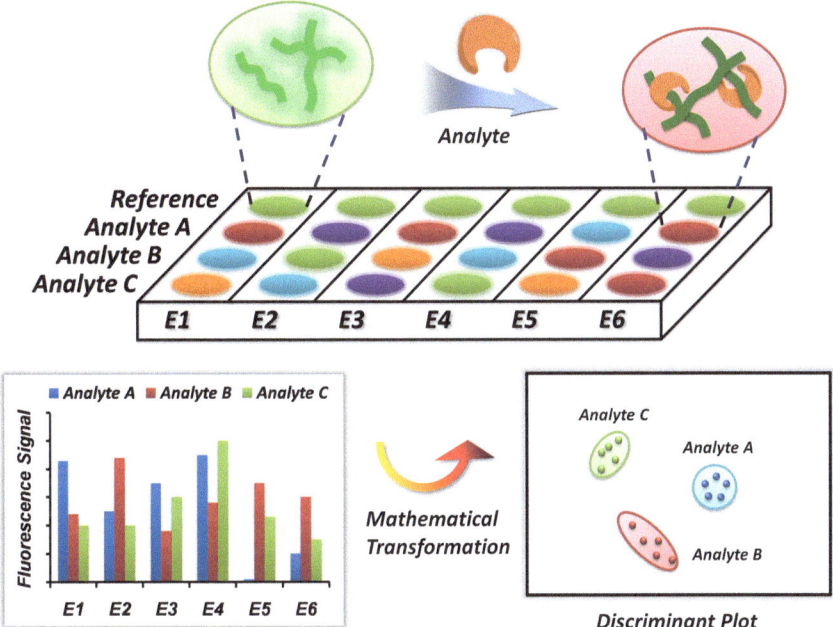

**Fig. 1.11** Basic concept of "chemical nose/tongue" strategy

into more intuitive discriminant plots. Moreover, upon constructing the standard discriminant plots, one can effectively assign other unknown samples by comparing the resultant response signals. The processes of sensing and data analysis are schematically depicted in Fig. 1.11, which have been successfully applied for diagnostics of microbial infection and discriminate tumors.

Owing to the superior properties of CPEs, such as high brightness, large extinction coefficients, superior photostability, large two photo (2P) action cross sections, low cytotoxicity, and versatile surface modification, fluorescence imaging with CPEs in vitro and ex/in vivo have also been successfully achieved. By virtue of the self-luminous property, CPEs can be employed as gene carriers for real-time tracking the location of gene delivery and transfection. CPEs were reported to be induced to generate ROS upon light illumination, which opened up a promising opportunity for new application of CPEs in therapeutics.

# References

1. Heeger AJ (2001) Semiconducting and metallic polymers: the fourth generation of polymeric materials (nobel lecture). Angew Chem Int Ed 40:2591–2611
2. Bernius MT, Inbasekaran M, O'Brien J, Wu WS (2000) Progress with light-emitting polymers. Adv Mater 12:1737–1750
3. Gather MC, Koehnen A, Meerholz K (2011) White organic light-emitting diodes. Adv Mater 23:233–248
4. Brabec CJ, Sariciftci NS, Hummelen JC (2001) Plastic solar cells. Adv Funct Mater 11:15–26
5. Dennler G, Scharber MC, Brabec CJ (2009) Polymer-fullerene bulk-heterojunction solar cells. Adv Mater 21:1323–1338
6. Dimitrakopoulos CD, Malenfant PRL (2002) Organic thin film transistors for large area electronics. Adv Mater 14:99–117
7. Li Y, Zou Y (2008) Conjugated polymer photovoltaic materials with broad absorption band and high charge carrier mobility. Adv Mater 20:2952–2958
8. McQuade DT, Pullen AE, Swager TM (2000) Conjugated polymer-based chemical sensors. Chem Rev 100:2537–2574
9. Thomas SW, Joly GD, Swager TM (2007) Chemical sensors based on amplifying fluorescent conjugated polymers. Chem Rev 107:1339–1386
10. Achyuthan KE, Bergstedt TS, Chen L, Jones RM, Kumaraswamy S, Kushon SA, Ley KD, Lu L, McBranch D, Mukundan H, Rininsland F, Shi X, Xia W, Whitten DG (2005) Fluorescence superquenching of conjugated polyelectrolytes: applications for biosensing and drug discovery. J Mater Chem 15:2648–2656
11. Feng X, Liu L, Wang S, Zhu D (2010) Water-soluble fluorescent conjugated polymers and their interactions with biomacromolecules for sensitive biosensors. Chem Soc Rev 39:2411–2419
12. Ho HA, Najari A, Leclerc M (2008) Optical detection of DNA and proteins moth cationic polythiophenes. Acc Chem Res 41:168–178
13. Bunz UHF (2000) Poly(aryleneethynylene)s: syntheses, properties, structures, and applications. Chem Rev 100:1605–1644
14. Swager TM (1998) The molecular wire approach to sensory signal amplification. Acc Chem Res 31:201–207
15. Zhu C, Liu L, Yang Q, Lv F, Wang S (2012) Water-soluble conjugated polymers for imaging, diagnosis, and therapy. Chem Rev 112:4687–4735

16. Liu B, Wang S, Bazan GC, Mikhailovsky A (2003) Shape-adaptable water-soluble conjugated polymers. J Am Chem Soc 125:13306–13307
17. Zhu C, Yang Q, Liu L, Wang S (2011) A potent fluorescent probe for the detection of cell apoptosis. Chem Commun 47:5524–5526
18. Moon JH, McDaniel W, MacLean P, Hancock LF (2007) Live-cell-permeable poly(p-phenylene ethynylene). Angew Chem Int Ed 46:8223–8225
19. Tang H, Duan X, Feng X, Liu L, Wang S, Li Y, Zhu D (2009) Fluorescent DNA-poly(phenylenevinylene) hybrid hydrogels for monitoring drug release. Chem Commun 6:641–643
20. Kim J-M, Lee J-S, Choi H, Sohn D, Ahn DJ (2005) Rational design and in situ ftir analyses of colorimetrically reversibe polydiacetylene supramolecules. Macromolecules 38:9366–9376
21. Ho H-A, Boissinot M, Bergeron MG, Corbeil G, Doré K, Boudreau D, Leclerc M (2002) Colorimetric and fluorometric detection of nucleic acids using cationic polythiophene derivatives. Angew Chem Int Ed 41:1548–1551
22. Phillips RL, Kim I-B, Tolbert LM, Bunz UHF (2008) Fluorescence self-quenching of a mannosylated poly(p-phenyleneethynylene) induced by concanavalin a. J Am Chem Soc 130:6952–6954
23. Phillips RL, Kim I-B, Carson BE, Tidbeck BR, Bai Y, Lowary TL, Tolbert LM, Bunz UHF (2008) Sugar-substituted poly(p-phenyleneethynylene)s: sensitivity enhancement toward lectins and bacteria. Macromolecules 41:7316–7320
24. Helms B, Mynar JL, Hawker CJ, Fréchet JMJ (2004) Dendronized linear polymers via "click chemistry". J Am Chem Soc 126:15020–15021
25. Englert BC, Bakbak S, Bunz UHF (2005) Click chemistry as a powerful tool for the construction of functional poly(p-phenyleneethynylene)s: comparison of pre- and postfunctionalization schemes. Macromolecules 38:5868–5877
26. Kim I-B, Shin H, Garcia AJ, Bunz UHF (2007) Use of a folate—PPE conjugate to image cancer cells in vitro. Bioconjugate Chem. 18:815–820
27. Lee K, Lee J, Jeong EJ, Kronk A, Elenitoba-Johnson KSJ, Lim MS, Kim J (2012) Conjugated polyelectrolyte-antibody hybrid materials for highly fluorescent live cell-imaging. Adv Mater 24:2479–2484
28. Pu K-Y, Li K, Liu B (2010) Multicolor conjugate polyelectrolyte/peptide complexes as self-assembled nanoparticles for receptor-targeted cellular imaging. Chem Mater 22:6736–6741
29. Chong H, Nie C, Zhu C, Yang Q, Liu L, Lv F, Wang S (2011) Conjugated polymer nanoparticles for light-activated anticancer and antibacterial activity with imaging capability. Langmuir 28:2091–2098
30. Liu B, Bazan GC (2004) Interpolyelectrolyte complexes of conjugated copolymers and DNA:platforms for multicolor biosensors. J Am Chem Soc 126:1942–1943
31. An L, Tang Y, Feng F, He F, Wang S (2007) Water-soluble conjugated polymers for continuous and sensitive fluorescence assays for phosphatase and peptidase. J Mater Chem 17:4147–4152
32. Li K, Zhan R, Feng S-S, Liu B (2011) Conjugated polymer loaded nanospheres with surface functionalization for simultaneous discrimination of different live cancer cells under single wavelength excitation. Anal Chem 83:2125–2132
33. Satrijo A, Swager TM (2007) Anthryl-doped conjugated polyelectrolytes as aggregation-based sensors for nonquenching multicationic analytes. J Am Chem Soc 129:16020–16028
34. Fang Z, Pu K-Y, Liu B (2008) Asymmetric fluorescence quenching of dual-emissive porphyrin-containing conjugated polyelectrolytes for naked-eye mercury ion detection. Macromolecules 41:8380–8387
35. Tomalia DA, Durst HD (1993) Genealogically directed synthesis—starburst cascade dendrimers and hyperbranched structures. Top Curr Chem 165:193–313
36. Pu K-Y, Li K, Shi J, Liu B (2009) Fluorescent single-molecular core—shell nanospheres of hyperbranched conjugated polyelectrolyte for live-cell imaging. Chem Mater 21:3816–3822

37. Yu M, Tang Y, He F, Wang S, Zheng D, Li Y, Zhu D (2006) Synthesis of water-soluble dendritic conjugated polymers for fluorescent DNA assays. Macromol Rapid Commun 27:1739–1745
38. Ahn DJ, Kim J-M (2008) Fluorogenic polydiacetylene supramolecules: immobilization, micropatterning, and application to label-free chemosensors. Acc Chem Res 41:805–816
39. Leclerc M (1999) Optical and electrochemical transducers based on functionalized conjugated polymers. Adv Mater 11:1491–1498
40. Ho HA, Dore K, Boissinot M, Bergeron MG, Tanguay RM, Boudreau D, Leclerc M (2005) Direct molecular detection of nucleic acids by fluorescence signal amplification. J Am Chem Soc 127:12673–12676
41. Aberem MB, Najari A, Ho H-A, Gravel J-F, Nobert P, Boudreau D, Leclerc M (2006) Protein detecting arrays based on cationic polythiophene-DNA-aptamer complexes. Adv Mater 18:2703–2707
42. Kushon SA, Ley KD, Bradford K, Jones RM, McBranch D, Whitten D (2002) Detection of DNA hybridization via fluorescent polymer superquenching. Langmuir 18:7245–7249
43. Kushon SA, Bradford K, Marin V, Suhrada C, Armitage BA, McBranch D, Whitten D (2003) Detection of single nucleotide mismatches via fluorescent polymer superquenching. Langmuir 19:6456–6464
44. Yang CJ, Pinto M, Schanze K, Tan W (2005) Direct synthesis of an oligonucleotide–poly(phenylene ethynylene) conjugate with a precise one-to-one molecular ratio. Angew Chem Int Ed 44:2572–2576
45. Tan C, Atas E, Müller JG, Pinto MR, Kleiman VD, Schanze KS (2004) Amplified quenching of a conjugated polyelectrolyte by cyanine dyes. J Am Chem Soc 126:13685–13694
46. Gaylord BS, Heeger AJ, Bazan GC (2002) DNA detection using water-soluble conjugated polymers and peptide nucleic acid probes. Proc Natl Acad Sci USA 99:10954–10957
47. Wang S, Gaylord BS, Bazan GC (2004) Fluorescein provides a resonance gate for fret from conjugated polymers to DNA intercalated dyes. J Am Chem Soc 126:5446–5451
48. Gaylord B, Heeger A, Bazan G (2002) DNA detection using water-soluble conjugated polymers and peptide nucleic acid probes. Proc Natl Acad Sci USA 99:10954–10957
49. Gaylord BS, Heeger AJ, Bazan GC (2003) DNA hybridization detection with water-soluble conjugated polymers and chromophore-labeled single-stranded DNA. J Am Chem Soc 125:896–900
50. Liu B, Bazan GC (2004) Homogeneous fluorescence-based DNA detection with water-soluble conjugated polymers. Chem Mater 16:4467–4476
51. Albert KJ, Lewis NS, Schauer CL, Sotzing GA, Stitzel SE, Vaid TP, Walt DR (2000) Cross-reactive chemical sensor arrays. Chem Rev 100:2595–2626

# Chapter 2
# Diagnostic Applications of Functionalized Conjugated Polyelectrolytes

**Abstract** In this chapter, the diagnostic applications of CPEs are illustrated from the following three aspects which are diagnostic sensors for disease-related biomarkers, diagnostics of microbial infection, and diagnostics of tumor. Förster resonance energy transfer (FRET) between PFP and fluorescein has been successfully utilized for detection of disease-related genes biomarkers, including DNA single nucleotide polymorphisms (SNPs), methylation, and mutation. The specific conformation-sensitive property of PDA and PT has been utilized to develop diagnostic methods toward bacteria and cancer cells. By taking advantages of signal amplification effect offered by CPEs, the concept of chemical nose/tongue has been employed for the detection and discrimination of proteins, bacteria, and discrimination between normal and cancer cells.

**Keywords** Diagnostic sensors for disease • Förster resonance energy transfer • Single nucleotide polymorphisms • DNA methylation • Diagnostics of microbial infection • Chemical nose sensor array • Tumor detection

## 2.1 Diagnostic Sensors for Disease-Related Biomarkers

The development of methods for DNA detection is of importance in disease diagnosis, gene-targeted drug discovery and molecular biology field [1–4]. Molecular diagnosis of disease-related biomarkers (e.g., DNA single nucleotide polymorphisms (SNPs), methylation, and mutation) with CPEs has evolved from laboratory studies to practical clinical applications. Besides DNA biomarkers, molecular diagnosis of some disease-related peptides or proteins (e.g., tumor markers) with CPEs has attracted much interest for the early detection of disease.

Facilitated by the perfect combination of cationic PFP 5 and fluorescein, numerous works have been conducted toward various nucleic acid-related detections. Modified from the assay devised by Bazan and co-workers [5, 6], Al Attar et al.

developed a PFP/surfactant/peptide nucleic acid (PNA)-based assay to achieve the effect of enhanced detection of drug-resistant mutants of ABL portion in the BCR-ABL oncogene [7]. FRET between PFP and Fl-labeled PNA/DNA distinguish fully complementary (mutant) and noncomplementary (single- and five-base mismatches) DNA targets. Different from previous assays, nonionic surfactants were introduced into the detection system, and the objective was to increase the fluorescence QY of PFP and reduce the nonspecific interaction between the PFP donor and the dye-labeled PNA acceptor in order to provide the system with a more sensitive feature.

SNPs represent a natural genetic variability found at high density in the human genome. SNPs are commonly recognized as genetic markers for mapping genes, defining population structure and performing gene association studies, as well as a fundamental tool in drug discovery and identification of genetic and inherited diseases [8–10]. CPE-based SNP detection has been developed by using similar sensing system as shown in Fig. 1.10 with the help of S1 nuclease enzyme [6]. The PNA-Fl probe is complementary to a region of the gene encoding the micro-tubule associated protein Tau. The probe sequence covers a known point mutation implicated in a dominant neurodegenerative dementia known as *FTDP-17*, which has many clinical and molecular similarities to Alzheimer's disease (AD). After specific hybridization with wild-type human DNA sequences or sequences containing a single-base mutation, the S1 nuclease enzyme is added to digest ssDNA sequences that are not perfectly hybridized with the PNA probe. Higher FRET signal is observed for the wild-type DNA compared with the mutant sequence harboring the SNP.

A label-free method has been developed to analyze SNP with CPE-amplified intercalating emitters (thiazole orange, TO) [11]. TO intercalates into the complementary stacked bases, and compared with direct excitation of TO, adding PFP leads to a 19-fold enhanced TO emission through CPE-excitation and FRET. The S1 nuclease can effectively degrade mismatched DNA/PNA complexes even for one-base mismatch, which directly results in the failure of TO intercalation.

SNP detection has also been achieved using the single-base extension (SBE) reaction [12], as illustrated in Fig. 2.1. The target DNA fragment is part of p53 exon8 containing a polymorphic site (G in wild-type target is replaced by A in the mutant). The $3'$-terminal T base of the probe is complementary to mutant-type target sequence instead of that for the wild-type target. Fluorescein-labeled G base (dGTP-Fl) can be incorporated into the mutant probe by an extension reaction, whereas for wild-type target the base extension reaction is blocked. On addition of PFP 5 into the extension systems, efficient FRET from PFP to fluorescein occurs for the mutant because of the closer distance between optical partners induced by electrostatic interactions. The method has high sensitivity and as low as 2 % allele frequency can be detected, which is shown in Fig. 2.1. The modified sensing platform, where dsDNA is used as a template instead of ssDNA, has been employed to discriminate the SNP genotypes of 76 individuals of Chinese ancestry [13]. Replacing the SBE reaction by incorporating dGTP-Fl and dUTP-Fl during the allele-specific PCR extension, one-step SNP detection has been achieved, which can be used to characterize genomic DNA [14]. Along these

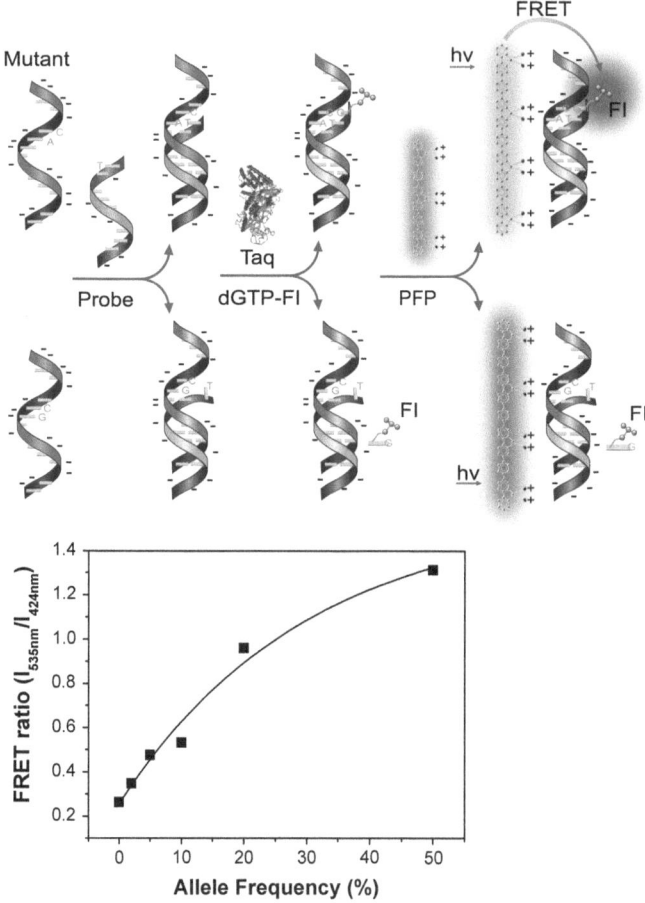

**Fig. 2.1** Schematic representation of DNA SNP detection mechanism based on CPEs and SBE reaction (*up*). FRET ratio ($I_{535nm}/I_{424nm}$) as a function of allele frequencies (*down*) (Reprinted with permission from Ref. [12]. Copyright (2007) American Chemical Society)

lines, a one-tube multicolor SNP genotyping assays method has been developed by means of PFP/DNA assemblies with multistep FRET between PFP and dyes incorporated in PCR extension products [15]. By analyzing FRET signals, the homozygous and heterozygous genotypes of SNPs were effectively discriminated in a single tube.

Leclerc and co-workers extended the well-developed homogenous PT/DNA sensory method to solid platforms to achieve ultrasensitive detection of SNPs. In 2006, on the basis of a solid DNA array composed of PT 23 and Cy3 labeled ssDNA FRET complexes, they realized the ultrasensitive discrimination of DNA targets (from a conserved region of *C. albicans* genome) between complementary and one-base mismatch ssDNA molecules with an LOD of 300 ssDNA copies for

**Fig. 2.2** Chemical structure of PFVP

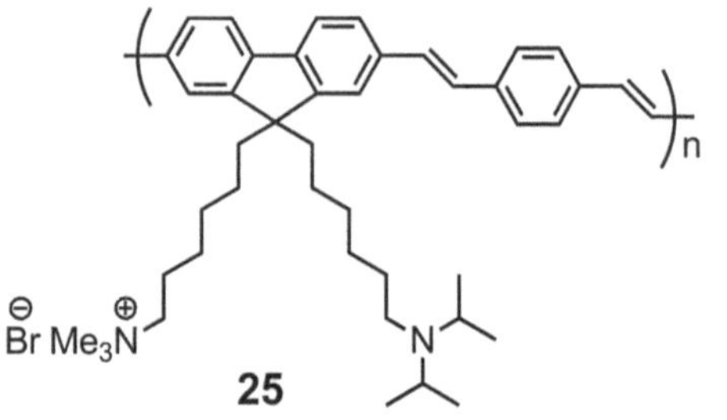

Br⊖ Me₃N⊕

**25**

the perfect integrate assays with a commercial laser source (408 nm) and further improve the detection sensitivity [16].

A microarray consisting of PNA-mobilized polystyrene (PS) beads was constructed for the detection of SNP by using cationic poly(fluorenyldivinylene-alt-1,4-phenylene) PFVP 25 (chemical structure shown in Fig. 2.2) [17]. The emission spectrum of PFVP overlaps with the absorption spectrum of Cy5, ensuring the possibility for FRET to occur from PFVP to Cy5. The fabrication of the array was self-assemble PS beads (500 nm in diameter) into the microwells on silicon substrates, followed by linking $NH_2$-terminated PNA probes to the NHS modified PS spheres. Specific hybridization of Cy5-labeled complementary ssDNA targets with PNA probes increased the negative charge density of PS beads, which promoted the cationic PFVP to keep in close proximity with Cy5 to favor the FRET process. For the specific Cy5-labeled DNA targets, the signals from the sensitized Cy5 could be detected as low as $10^{-17}$ M by CLSM. By taking 300 µL of the DNA target into consideration, the absolute amount was calculated to be at the zeptomole level. The ultrahigh sensitivity was attributed to the large surface area provided by PS beads and the light-amplification property of PFVP. In contrast, even when the concentration of the single-base mismatch ssDNA reached up to $10^{-8}$ M, no signals were detected because of the extremely unstable character of mismatched PNA/DNA duplexes, exhibiting the high selectivity. By employing unlabeled ssDNA, the assay could also be broadened for label-free detection of DNA by directly monitoring the fluorescence emission of PFVP.

DNA methylation, an important component of epigenetic regulation, describes the incorporation of a methyl group to the 5-position of the cytosine pyrimidine ring or the number 6 nitrogen of the adenine purine ring [18–20]. Hypermethylation commonly occurs at CpG islands (clustered CpGs sites) in the promoter region of tumor suppressor gene and subsequently suppresses the transcription of the corresponding gene, ultimately leading to aberrant protein expression and in cancer formation and progression. Building on the principle of SNP detection, a protocol for site-specific CpG methylation detection has been established [21]. In this approach, bisulfite treatment changes nonmethylated C into U but does not influence the methylated C, and the resulting U is substituted by T after PCR amplification [22]. After SBE, dGTP-Fl is incorporated into the probe for the methylated DNA, but not for nonmethylated DNA (Fig. 2.3). On addition of PFP 5, efficient FRET from PFP to fluorescein occurs for the methylated DNA

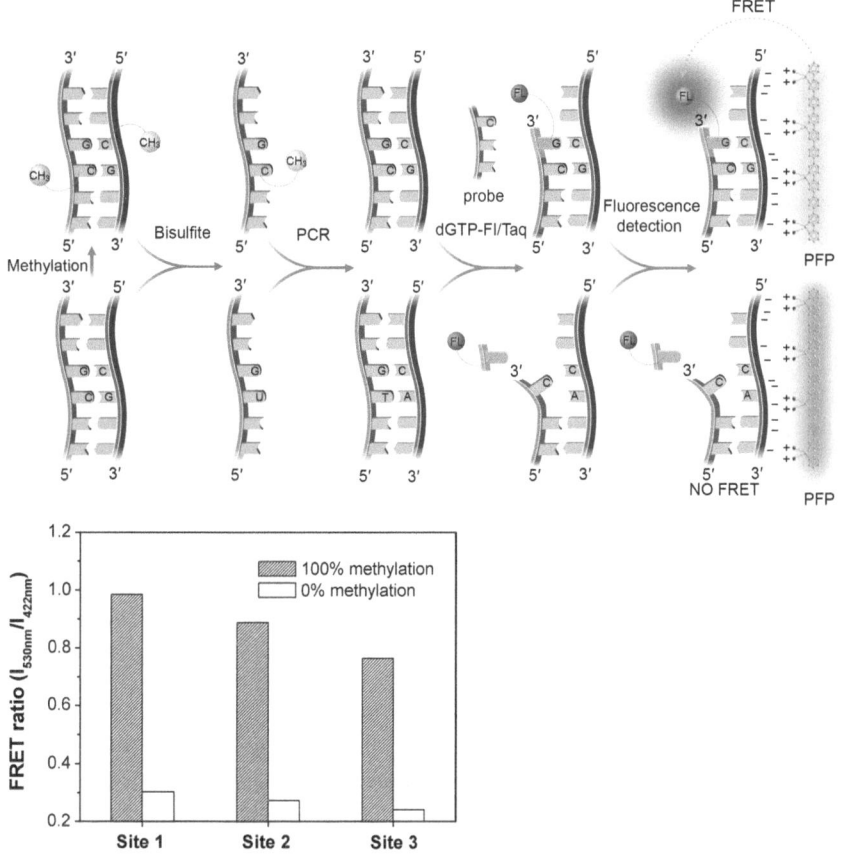

**Fig. 2.3** Schematic representation of DNA methylation detection (*up*). FRET ratio ($I_{530nm}/I_{422nm}$) of the extension products with methylated DNA and nonmethylated DNA in the presence of PFP 5 (*down*) (Reprinted with permission from Ref. [21]. Copyright (2008) American Chemical Society)

but not for nonmethylated DNA. By virtue of this protocol, the methylation status of three CpG sites in the p16 promoter region of the human colon cancer cell line HT29 has been detected with high sensitivity, and the results are shown in Fig. 2.3. The FRET ratios ($I_{530nm}/I_{422nm}$) for specific extension are 3-5 times higher than that of the nonspecific extension.

A protocol for site-specific CpG methylation detection has been established by Wang's group [23]. In this protocol, a methylation-sensitive restriction endonuclease (HpaII) was introduced to specifically digest unmethylated recognition sites that were directly derived from a small amount of genomic DNA, while leaving the methylated DNA intact. Subsequent nested PCR amplification will exclusively incorporate Fl-dNTPs into the PCR products for methylated DNA but not for unmethylated one. Upon addition of polymer 5, distinct FRET from polymer 5 to

Fl was observed for situation B. The reported study also investigated the methylation statuses of three cancer suppressor genes (p16, HPP1, and GALR2 promoters) from five cancer cell lines, HT29, HepG2, A498, HL60, and M17. These results afford extremely significant correlation information between cancers and susceptibility genes, which is very useful for early cancer diagnosis.

Compared with single methylation alteration, assessing combined methylation alterations can provide higher association with specific cancer. DNA methylation levels of seven colon cancer-related genes in a Chinese population have been quantitatively analyzed on the basis of FRET from PFP 5 to dyes [24]. Through a stepwise discriminant analysis and cumulative detection of methylation alterations, high accuracy and sensitivity for colon cancer detection (86.3 and 86.7 %) and for differential diagnosis (97.5 and 94 %) were acquired. Moreover, a correlation between the CpG island methylator phenotype and clinically important parameters in patients with colon cancer was identified. The cumulative analysis of promoter methylation alterations by the CPEs-based FRET may be useful for the screening and differential diagnosis of patients with colon cancer, and for performing clinical correlation analyses.

In addition to SNP and methylation detection, other alterations with respect to DNA, such as DNA mutation and DNA lesion, have also been investigated. The L858R mutation of epidermal growth factor receptor (EGFR) in non-small cell lung cancer is associated with the increased sensitivity to EGFR tyrosine kinase inhibitors. Wang and co-workers developed a simple and sensitive method for identification of L858R mutation in cell lines and tumor tissues using CPEs-based FRET method [25]. The detection system could detect as low as 4–8 % mutation of the total DNA. Through the detection results for 48 DNA samples from tumor tissues, a sensitivity of 95.24 % (20/21) and a specificity of 96.30 % (26/27) were demonstrated. Further, the application of this method in clinical molecular diagnosis was validated by detecting T790Min EGFR of 35 patients. Particularly, this method could confirm the suspected positive samples arisen by DNA sequencing and real-time PCR methods.

On the basis of multistep FRET of CPEs, a visual colorimetric method for detecting multiplex DNA mutations has been developed using multicolor fluorescent coding. The visual system provides a quantitative detection by simply analyzing RGB values of images [26]. Genomic DNAs extracted from formalin-fixed paraffin-embedded (FFPE) colon tissues can be sensitively determined by utilizing visual assay with a high-throughput manner (Fig. 2.4).

Aneja et al. studied DNA mutation detection of *B. subtilis* bacteria using a FRET-based method [27]. The perfect hybridization between Fl-PNA and ssDNA target can mediate efficient FRET from PFP to Fl via electrostatic attraction. With the increase of mutation number, the FRET efficiency gradually decreased with sufficient sensitivity to detect up to 4 base mismatches.

Besides DNA sequence alterations, other mutation types can also be analyzed by using CPEs. In 2009, He and co-workers utilized aldehyde-containing cationic PT copolymer 26 (chemical structure shown in Fig. 2.5) to successfully realize the detection of dsDNA deletion mutation [28]. Initially, addition of probe ssDNA

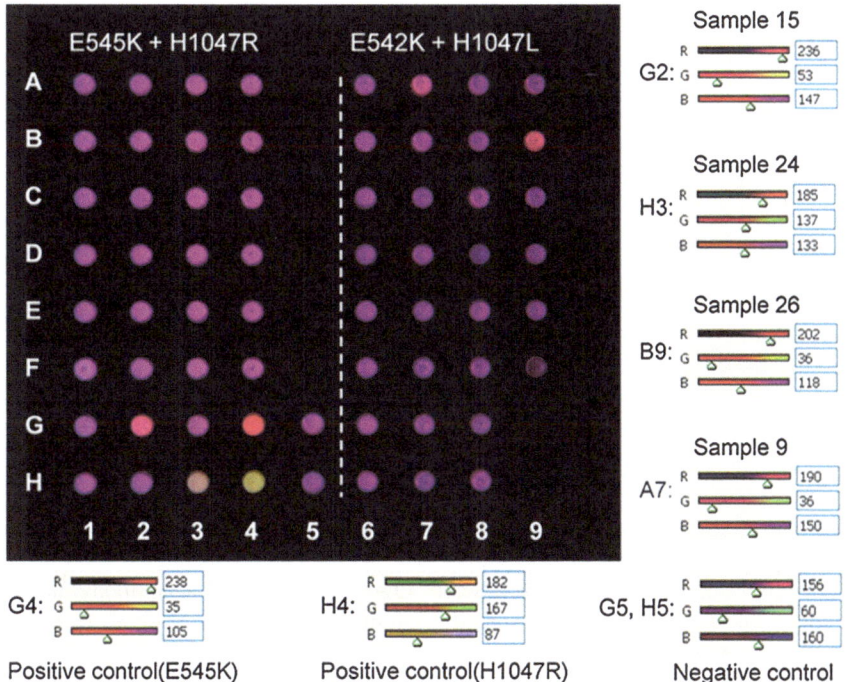

**Fig. 2.4** Visual detection of 30 colon adenocarcinomas samples. On the *left side* of the *white dotted line*, E545K and H1047R primers were used to detect E545K and H1047R mutation sites. G4 and H4 are two positive controls (MCF-7 cell lines and T47D cell lines); G5 and H5 are two negative controls (293T cell lines). On the *right side* of the *dotted line*, E542K and H1047L primers were used to detect E542K and H1047L mutation sites. The 30 colon adenocarcinomas samples are identical to the *left of dotted line* (Reprinted with permission from Ref [26]. Copyright (2012) American Chemical Society)

could dramatically quench the fluorescence of PT 26 via electronic/energy transfer. In the presence of target ssDNA, the system fluorescence could recover to different extend by forming a triple complex. The authors attributed the different responses to the function of aldehyde groups in the backbone of PT and speculated that subtle differences were magnified via hydrogen bonds and chemical binding interactions between aldehyde groups and the unpaired bases in the mutant DNA. In the concentration range of $0.5–25 \times 10^{-9}$ M, this method afforded linearity, and a real sample test (bird flu-type H1N1 oligonucleotides) was performed in an acceptable way. Supplementary to the mutation detection, Wang and co-workers developed a DNA lesion detection method by using the prominent PFP 5/Fl FRET pair [29]. Cyclobutane pyrimidine dimer (CPD) and pyrimidine-pyrimidone dimer (PPD) can be induced when two adjacent T bases are irradiated with UV light, which causes DNA lesions. In this case, SBE of Fl-dUTP was restricted, whereas the incorporation of Fl into the extended DNA chain would take place for the intact DNA template. After the addition of PFP 5 into the system, weak or

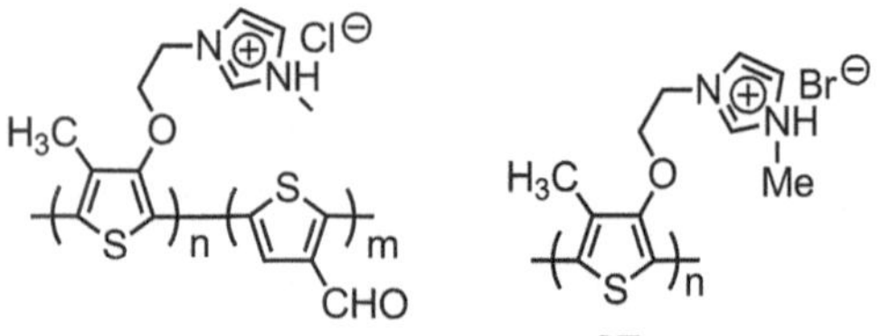

**Fig. 2.5** Chemical structures
of CPEs 26 and 27

**26**          **27**

no FRET was detected for the damaged DNA; but for the intact DNA, one could observe an obvious FRET signal. The total test time was between 30 min and 1 h, and the detection limit of damaged DNA was below a nanomolar level.

Besides DNA biomarkers, disease-related peptides or proteins (e.g., tumor markers) are equally important in molecular diagnosis. Sim and co-workers reported the detection of prostate cancers based on PDA supramolecules by a hybrid stimulus amplified method [30]. In view of the clinical importance of prostate cancers, prostate-specific antigen-$\alpha$1-antichymotrypsin (PSA-ACT) complex was selected as the tumor marker. The antibody-functionalized cross-linked PDA chip was successfully constructed. Owing to the specific antibody-antigen recognition, the PDA chip presented a nonfluorescent to fluorescent transition upon the addition of PSA-ACT, and the LOD was 10 ng·mL$^{-1}$. However, in contrast with cells, the response signals to proteins were relatively weak due to their smaller sizes. To further improve the detection sensitivity in order to meet the clinic requirement (<4 ng·mL$^{-1}$ for normal human PSA), secondary antibody-conjugated magnetic beads (linked via biotin-STA interaction) were utilized to strengthen the initial stimuli that resulted from the primary antibody-antigen reaction. Facilitated by the subsequent mechanical force from the magnetic beads, the LOD of this method toward PSA-ACT was enhanced to be 0.1 ng·mL$^{-1}$. Considering the reduced sample consumption and increased diagnostic reliability, simultaneous detection of multiple cancer markers should be preferred in practical clinical application. Utilizing cationic PT 27, multiplexed detection of cancer markers on the surfaces of Au/Ag barcoded nanorods was developed, where three cancer markers, prostate-specific antigen (PSA), carcinoembryonic antigen (CEA), and $\beta$-human chorionic gonadotropin ($\beta$hCG) were involved [31]. Immobilization of cancer markers onto the antibody-functionalized nanorods was realized by the specific immune recognition. Next, dsDNA-containing secondary antibody (aided by biotin-STA binding) was added to improve the negative charge density of the nanorod surface. Finally, electrostatic complexation between polymer 27 and negatively charged dsDNA led to the intense fluorescence emission of nanorods, achieving the aim of specific and sensitive detection of cancer antigens. By means of fluorescence technique, the LOD was determined to be 0.16 ng·mL$^{-1}$ for PSA, 0.21 ng·mL$^{-1}$ for CEA, and 0.08 ng·mL$^{-1}$ for $\beta$hCG, greatly less than the clinical thresholds (4, 2.5, and 3 ng·mL$^{-1}$, respectively). In addition to the advantage of sensitive fluorescence signals from PT, the encoding capability of nanorods themselves (three distinct striping patterns, 000100, 01010, and 011110) made this method potent enough to simultaneously detect various cancer markers by labeling the barcoded nanorods with different mAbs. In conjunction with the strong scattering ability of the nanorods, multiplexed detection of cancer markers was demonstrated. Moreover, multiplexed detection of cancer markers was also effective in bovine serum.

The detection of glutathione (GSH, with two negative charges) using cationic PT 28 in both colorimetric and fluorescent manner was demonstrated [32]. Through an in situ premodification method illustrated in Fig. 2.6, o-phthalaldehyde (OPA) reacted with two nucleophile-containing GSH (thiol and amine) to form the corresponding isoindole derivative to introduce the hydrophobic aromatic moiety, which was necessary to realize analyte-induced aggregation of PTs. Upon the addition of isoindole-GSH, the adsorption of polymer 28 red-shifted from 407 to 545 nm and the solution color changed from yellow to pink-red as a result of the aggregation of PTs from a random-coiled state to a chirally ordered state. The specificity and selectivity toward GSH over 20 amino acids and thiol-bearing molecules (homocysteine and bovine serum albumin) were demonstrated by monitoring the color change of PTs or by calculating the relative absorbance change (Fig. 2.6). Fluorometric detection of GSH was also feasible by recording the fluorescence of PT 28 at 557 nm, and the LOD was 10 nM.

A multicarboxylated PPE 29 was synthesized for sensitive fluorescence turn-on detection of glutathione reductase (GR) [33]. GR is an important enzyme to facilitate the conversion of oxidized glutathione (GSSG) to reduced GSH. Four carboxylate groups per repeat unit (RU) in PPE 29 (chemical structure shown in Fig. 2.7) impart water solubility and relatively high quantum yield (11 %) due to less aggregation in water. In fact, the pendant chemical structure of PPE was very

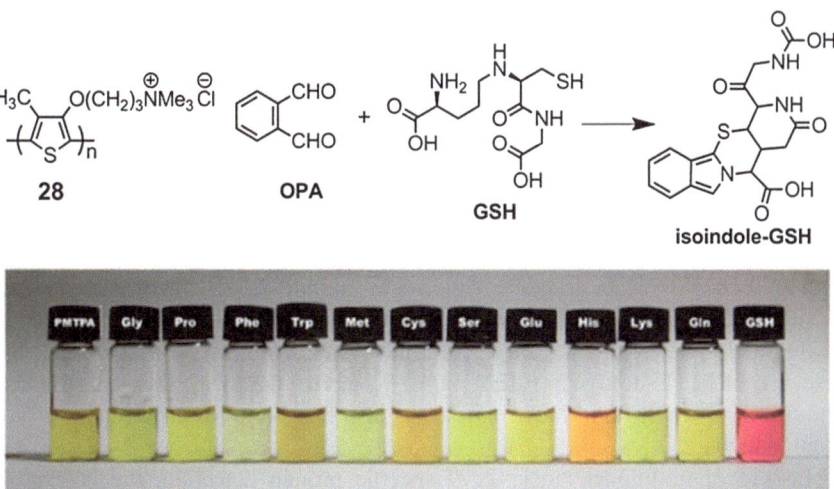

**Fig. 2.6** Schematic illustration of the in situ premodification reaction of GSH with OPA and the selective detection of GSH (Reprinted with permission from Ref. [32]. Copyright (2009) Royal Society of Chemistry)

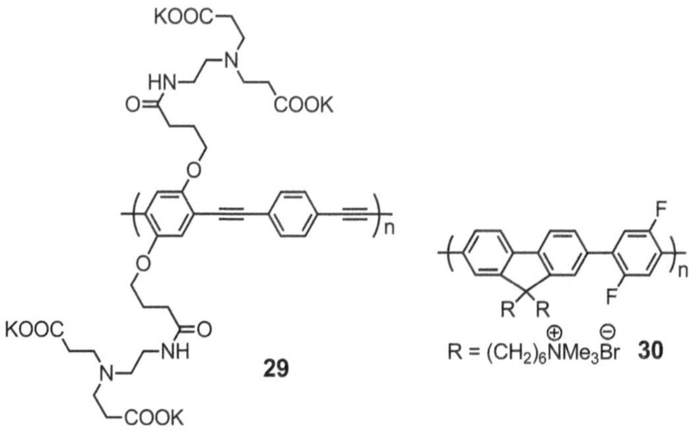

**Fig. 2.7** Chemical structure of PPE 29 and PFP-2F 30

similar to EDTA, exhibiting the ability to chelate metal ions. The dramatically quenched fluorescence of PPE 29 by $Cu^{2+}$ was obtained. Taking advantage of the selective response of PPE 29 to $Cu^{2+}$, the combination of PPE 29/$Cu^{2+}$/GSSH could be employed to detect GR in a fluorescence turn-on manner, which was built on the fact that GSH could efficiently displace $Cu^{2+}$ by forming stable complexes with the free thiol to resume the original quenched fluorescence. In the presence of coenzyme NADPH, GR activity could be tested by monitoring the fluorescence recovery of PPE. In this study, the fluorescence recovery ratio was more than 10-fold, and the LOD for GR was 0.2 mU·mL$^{-1}$. Both the inhibition effect by GR inhibitor and the selectivity for GR over other proteins were also demonstrated.

In terms of real-world applications, it is essential to develop an assay for the detection of important proteins in complicated biological media with high sensitivity and selectivity. Taking advantage of the light-amplification property of PFP-2F 30 (chemical structure shown in Fig. 2.7), Liu and co-workers reported the detection of thrombin in 10 % blood serum using aptamer-functionalized silica NPs [34]. The specific interactions between thrombin and the specific aptamer (thrombin binding aptamer (TBA), 15-mer, and 29-mer DNA) favored the formation of a fluorescent sandwich structure, which was composed of 29-mer primary TBA modified silica NPs, thrombin, and 15-mer Fl-labeled secondary TBA. The following addition of cationic PFP-2F into the system extremely enhanced the fluorescence of Fl (13-fold) via FRET. In 10 % serum, the assay was still effective, and the LOD was 1.06 nM. The assay was also feasible for other aptamer/protein combinations.

Instead of traditional "lock–key" specific protein detection, the concept of chemical nose/tongue was employed for protein detection and discrimination. Bunz and co-workers first utilized six gold NPs-PPE chemical nose sensors to realize the identification of various proteins [35]. Upon addition of proteins with diverse MW and pI, the quenched electrostatic complexes (gold NP-PPE) were disrupted via different interactions with proteins, leading to the differential displacement of anionic polymers so as to generate distinct fluorescence-response patterns. Six cationic gold NPs (NP1-NP6) with different hydrophobicities and hydrogen-bonding functionalities were utilized to regulate diverse interactions with seven proteins. Combined with LDA, the proteins could be effectively discriminated, even for the proteins with similar MW and pI. To eliminate the influence of varying protein concentration in practical samples, by referring to the preconstructed standard training matrix (6 NPs × 6 replicates × 7 proteins) where

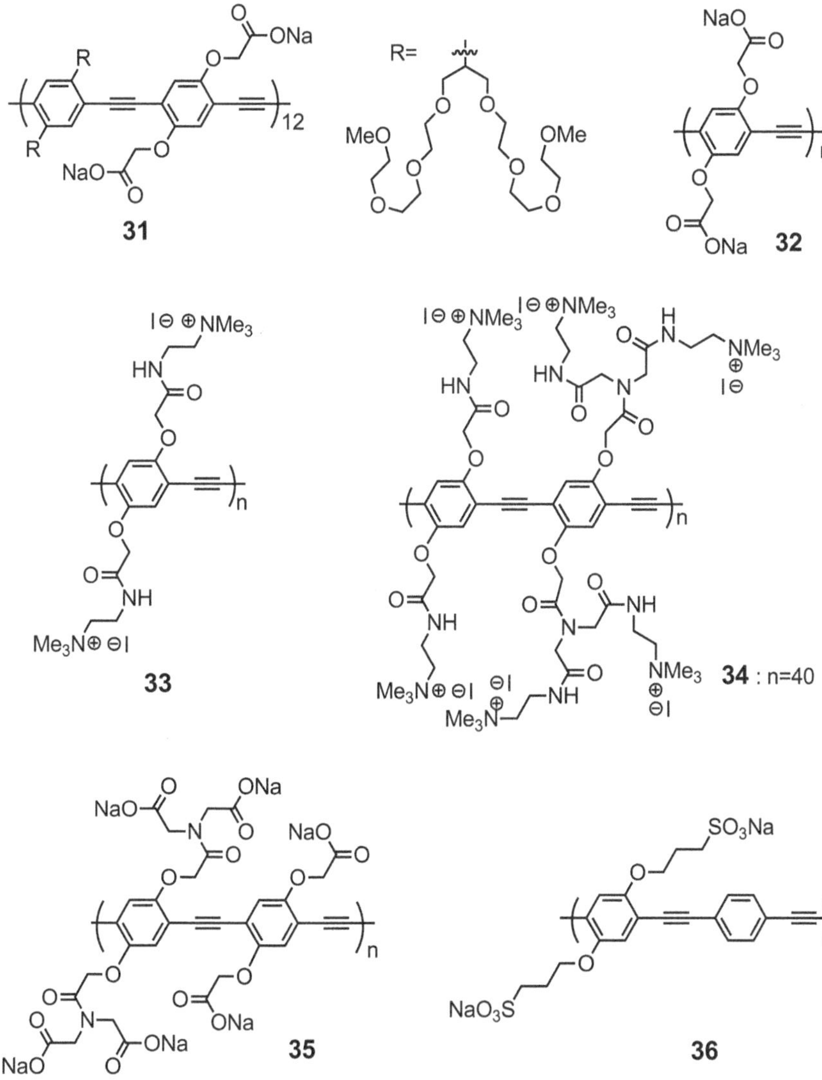

**Fig. 2.8** Chemical structures of CPEs 31 to 36

the protein concentration was fixed at $A_{280} = 0.005$ ($A_{280}$ represents the ultraviolet absorbance at 280 nm), 52 unknown protein samples from 7 different proteins were successfully assigned.

Subsequently, replacing the gold NP-PPE sensors by six functionalized PPEs (31–36, chemical structures shown in Fig. 2.8), Bunz and co-workers achieved the aim of protein detection and discrimination [36]. Six PPEs featured as different charge characteristics and molecular scales were used for presenting diverse binding abilities to discriminate 17 proteins (with different MW, pI, UV absorbance, and metal/nonmetal containing situations). The response patterns were defined as the net changes of fluorescence signals at 465 nm ($\Delta I$). With the aid of LDA, a discrimination accuracy of 100 % was obtained. As for 68 unknown samples from the 17 proteins, the assignment could be accomplished with an accuracy of 97 %.

Responsive elements for array detection can also be obtained by mixing the cationic conjugated oligoelectrolyte FPF 37 (chemical structure shown in Fig. 2.9) and Fl-ssDNA [37]. This combination leads to the formation of electrostatic

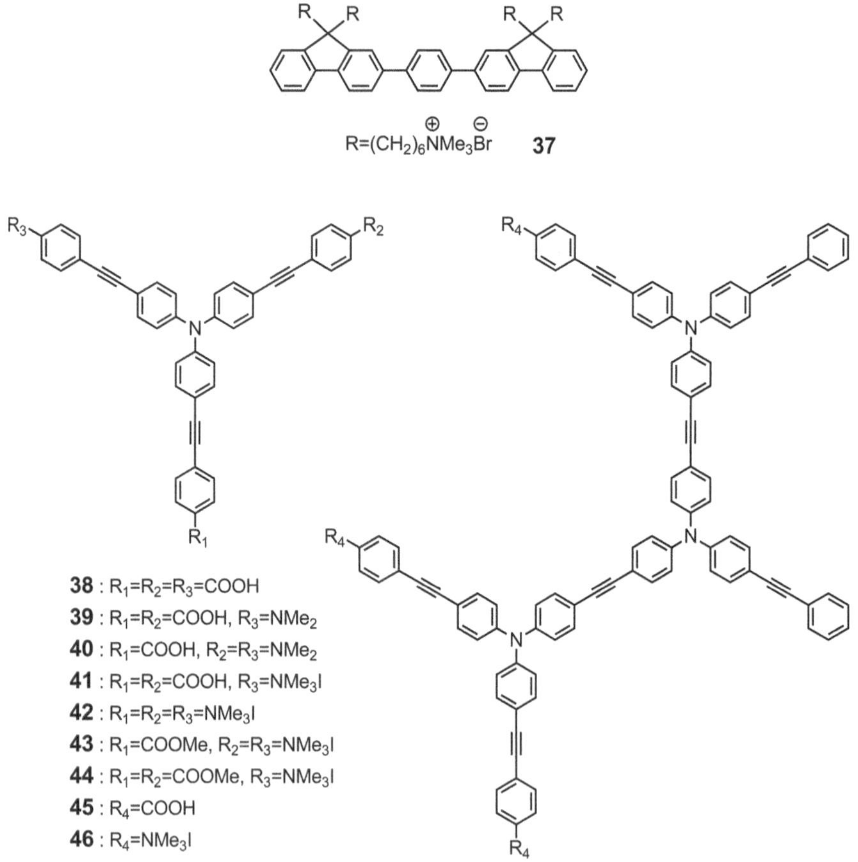

$$R = (CH_2)_6 \overset{\oplus}{N}Me_3 \overset{\ominus}{Br} \quad \mathbf{37}$$

**38** : $R_1 = R_2 = R_3 = COOH$
**39** : $R_1 = R_2 = COOH, R_3 = NMe_2$
**40** : $R_1 = COOH, R_2 = R_3 = NMe_2$
**41** : $R_1 = R_2 = COOH, R_3 = NMe_3I$
**42** : $R_1 = R_2 = R_3 = NMe_3I$
**43** : $R_1 = COOMe, R_2 = R_3 = NMe_3I$
**44** : $R_1 = R_2 = COOMe, R_3 = NMe_3I$
**45** : $R_4 = COOH$
**46** : $R_4 = NMe_3I$

**Fig. 2.9** Chemical structures of CPEs 37 to 46

aggregates. For the detection platform, five Fl-labeled ssDNA with different sequences and lengths were used. The addition of different proteins results in unique spectral responses due to the diverse hydrophobic and electrostatic interactions. The spectral changes can be quantitatively represented as two variables $\gamma_G$ and $\gamma_B$, which are the emission intensity ratios at 400 nm (FPF emission) and 525 nm (Fl emission) in the presence and absence of proteins. Transforming of the $\gamma_G$ and $\gamma_B$ response patterns by PCA leads to independent and mutually complementary score plots that identify different protein, even for three phosphorylated BSAs.

In chemical nose/tongue strategies, abundant components with variable structural characteristics are favorable to achieve ideal discrimination effects, and systematic regulation of surface charges is an efficient method for fine-tuning complicated supramolecular interactions. Making use of a series of dendritic phenylene-ethynylene fluorophores with different charge situations, Sukwattanasinitt and co-workers developed an array for protein discrimination [38]. By diverse combination of the substituent groups of cationic quaternary ammonium, anionic carboxyl group, and nonionic methyl ester or N,N-dimethylamino group, the authors synthesized nine dendritic fluorophores (38–46, seven zeroth generations and two first generations, chemical structures shown in Fig. 2.9) with varying charges. After the addition of proteins, only five fluorophores were induced to produce fluorescence changes, resulting in a 5 fluorophores × 8 proteins × 9 replicates matrix. PCA was then utilized to transform the raw response patterns (quantified as $\Delta I$) into eight separate clusters. Factorial discriminant analysis (FDA) cross-validation was used to quantify the classification accuracy, the optimal detection wavelength (500 nm) was determined. Moreover, through further optimization, the number of sensing fluorophores could be reduced from five to two while the discriminant accuracy was still maintained at 100 %.

Wang and co-workers designed cationic 9,9-bis(6'-bromohexyl)-2,7- dibromofluorene (PFDE) 47 (chemical structure shown in Fig. 2.10) for protein identification and denaturation detection [39]. The special design of diene-containing polymer was to increase the backbone flexibility for protein sensing. The discriminant array was the combination of PFDE under different ionic strengths (10–60 mM). Similar to the principle developed by Bunz and co-workers, proteins differentially interfered with PFDE through electrostatic and hydrophobic interactions to alter its fluorescence signals, generating diverse response patterns. Through LDA transformation, seven proteins were totally discriminated without any overlap, and the LOD was 0.1 µM. When proteins are subjected to denaturation, the morphology and aggregation states of proteins will undergo remarkable

**Fig. 2.10** Chemical structure of PFDE 47

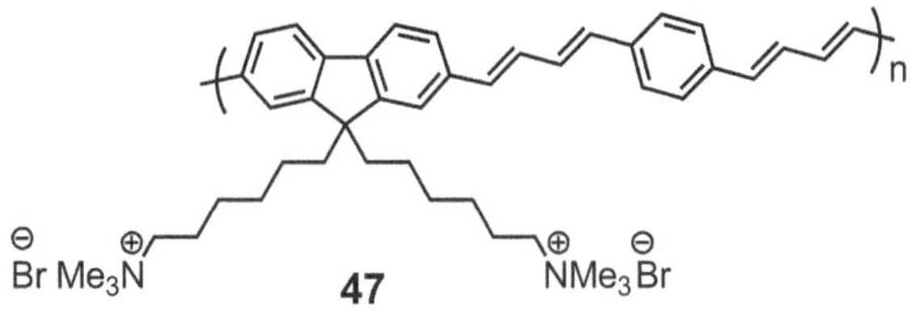

changes. Interestingly, PFDE 47 was potent to probe this process by exhibiting an obvious enhancement in the emission intensity for all seven proteins, which was explained by the reduced interchain $\pi-\pi$ interaction on the larger denatured protein aggregates.

Besides protein discrimination and detection, CPEs are also employed to identify protein misfolding. Nilsson and co-workers dedicated numerous contributions to the detection of amyloid fibril formation by using conformation-sensitive PTs [40, 41]. Many protein misfolding-related diseases, such as AD, systemic amyloidoses, and prion diseases, are featured as extracellular deposition of amyloid fibrils. Moreover, aggregated protein deposits with amyloid characteristics are also found in Huntington's disease. On the basis of their previous studies on the detection of fibrillar conformation of amyloidogenic proteins (insulin and lysozyme) by anionic poly(thiophene acetic acid) (PTAA) 48 (chemical structure shown in Fig. 2.11), they subsequently unfolded extensively systematic investigation on protein misfolding. The basic principle is that PTs change their conformation according to the secondary structures of intrinsically positively charged proteins, such as native α-helical and misfolding cross-β-sheet structures. By means of flow linear dichroism spectroscopy (flow-LD), Westerlund and co-workers demonstrated that, analogous to standard amyloid-specific dyes, PTs were bound in the grooves between adjacent protein side-chains in the amyloid fibril core, preferentially parallel to the fibril long axis [42]. Such a conformational change can lead to the variation of optical properties. When the protein exists in the form of a normal α-helix, the absorption and emission wavelengths of PTs are relatively blue-shifted, indicating the more helical status. However, in the presence of amyloid fibrillar proteins, PTs display red-shifted absorption and emission spectra as a result of the more planar conformation. Therefore, the conformation changes of amyloidogenic proteins could be determined by monitoring the changes of the absorption and emission spectra of PTs or simply visualization by naked eyes. In 2007, Nilsson and co-workers investigated the 2P absorption cross-section of PTs 49 (chemical structure shown in Fig. 2.11) and applied the most optimal PT for protein amyloid conformation detection [43]. Inganäs and co-workers recently investigated the kinetics of prefibrillar amyloid formation and molecular interactions of p-FTAA (50, chemical structure shown in Fig. 2.11) with bovine insulin [44]. Results demonstrated that,

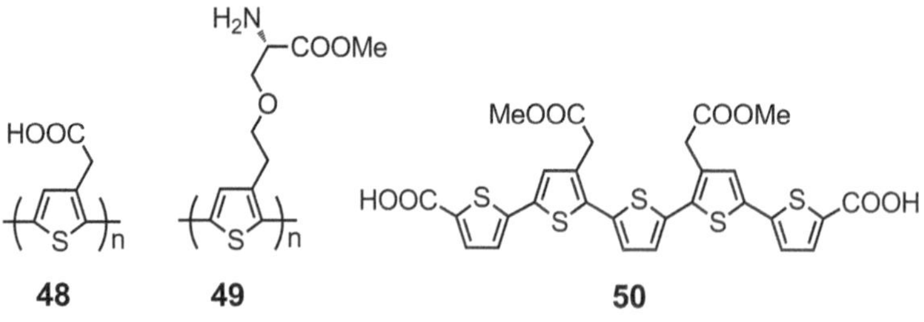

**Fig. 2.11**  Chemical structure of CPEs 48 to 50

prior to the formation of mature insulin amyloid fibrils, p-FTAA exhibited binding ability toward prefibrillar aggregates of insulin.

To realize clinical applications of CPEs, the well-developed CPEs-based detection system would be effective in biological fluids (e.g., blood and urine), which is still a challenge for CPEs and more efforts are needed to address this issue.

## 2.2 Diagnostics of Microbial Infection

Many pathogens that infect humans use cell-surface carbohydrates as receptors to facilitate cell–cell adhesion. The hallmark of these interactions is their multivalency, or the simultaneous occurrence of multiple interactions. Seeberger and coworkers reported the detection of the bacteria e*scherichia coli (E. coli)* by using carbohydrate-functionalized PPE [45]. Baek et al. reported the detection of pathogens (*E. coli* and influenza virus) using glycopolythiophenes pendant with sialic acid or mannose ligands [46]. The utilized principle was based on the absorption spectrum changes of the polymers. On the basis of identical principle, Bunz and co-workers developed two α-mannose-functionalized PPEs by postpolymerization functionalization for the detection of *E. coli*. The detection mechanism was analyte-induced aggregation. They increased linker length between the sugar and the backbone of PPEs to enhance the flexibility of side-chains and incorporated more complex trimannose to further improve the response sensitivity.

Liu and co-workers synthesized neutral sugar-containing (β-glucose and α-mannose) PFPs (51–53, chemical structures shown in Fig. 2.12) with different lengths of ethylene glycol spacers for the detection of *E. coli* [47]. In contrast to

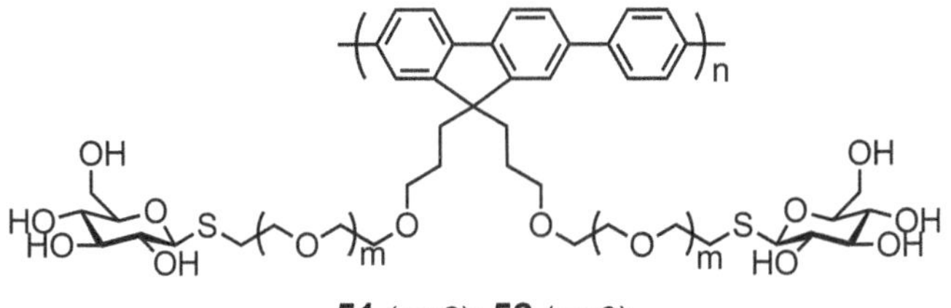

**51** (m=2); **52** (m=9)

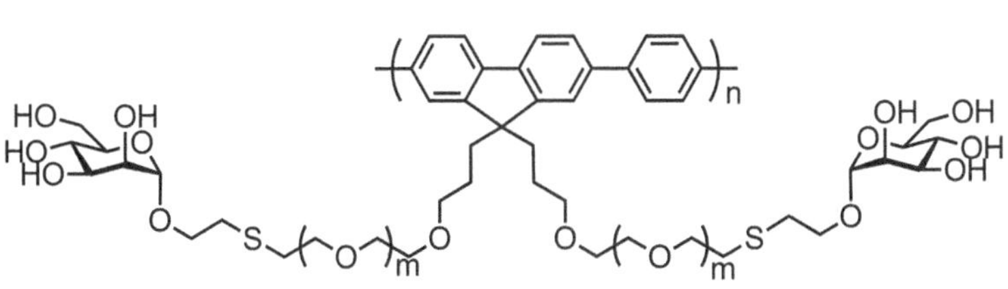

**53** (m=9)

**Fig. 2.12** Chemical structures of CPEs 51 to 53

polymers with a short oligo(ethylene glycol) (OEG) linker (i.e., polymer 51 with β-glucose), polymers with a long PEG spacer (polymer 52 with β-glucose and 53 with α-mannose) exhibited relatively good water solubility and simultaneously offered the sugar components greater flexibility. The polymers can form highly fluorescent aggregates in the presence of *E. coli* ORN 178, but not for the mannose nonbinding *E. coli* ORN 208. The direct evidence was the red-shifted UV/vis absorption and fluorescence emission spectra, which stemmed from the increased π-conjugation degree.

Inspired by the excellent work from Leclerc group for the detection of *C. albicans* [48], a label-free detection method for hepatitis B virus (HBV) oligonucleotides has been developed by He and co-workers using cationic PT 23. [49]. The yellow solution of cationic PT corresponds to short-wavelength absorption owing to the limited conjugation length. However, the addition of negatively charged capture oligonucleotide probes could transform the backbone of PT into a more planar conformation by forming duplex strands, and the solution color changed into red. In this case, the fluorescence of PT 23 was dramatically quenched due to the formation of planar duplex aggregates. However, in the presence of complementary oligonucleotides, the formative triplex composed of hybridized double strands and cationic PTs led to a 15-fold enhanced fluorescence. On the basis of a reversible fluorescence quenching mechanism, the system was used to specifically recognize oligonucleotides related to YMDD gene mutation of HBV. Only perfectly complementary targets (wild-type) could maximally recover the systematic fluorescence, and a distinctive response from 1-bp (base pair) mismatch analytes could be identified. In the range of 0.1–10 nmol·L$^{-1}$, a linear relationship was shown and the detection limits were 0.088 nmol·L$^{-1}$ for wild-type targets and 0.102 nmol·L$^{-1}$ for 1-bp mismatch targets, respectively.

PDA has been extensively studied for pathogen detection owing to their sensitive response ability. In 2007, Sim and co-workers developed a generic bioanalytical matrix with a liquid-phase antibody-functionalized PDA-sensing system [50]. The PDA liposomes with varying molar ratios of 10,12-pentacosadiynoic acid (PCDA) and a PCDA derivative with a maleimide headgroup (PCDA-MI) were prepared to investigate the influence of PCDA-MI contents on the sensitivity and stability of PDA biosensors. Results indicated that the membrane fluidity and colorimetric response of PDA linearly increased with the increase of PCDA-MI contents in the range of 0–30 %. Various thiol-bearing receptors (e.g., antibody) could be attached to the PDA liposomes by their reactions with reactive maleimide groups. Anti-*C. parvum* monoclonal antibody (mAb) was used for conjugation to afford liposome. Compared with control Bacillus spores, a colorimetric response of antibody-functionalized liposome to *C. parvum* was observed, which originated from the specific antibody-antigen interactions. As expected, the detection sensitivity increased with the increased antibody contents. Another PDA-based liquid-phase system was reported by Wang and co-workers for the specific detection of shiga toxin (Stx)-producing *E. coli* O157 by Gal-α1,4-Gal disaccharide functionalized PDA liposome 54 (chemical structure shown in Fig. 2.13) [51]. To overcome the relative instability of liquid-phase PDA sensors, Park and co-workers

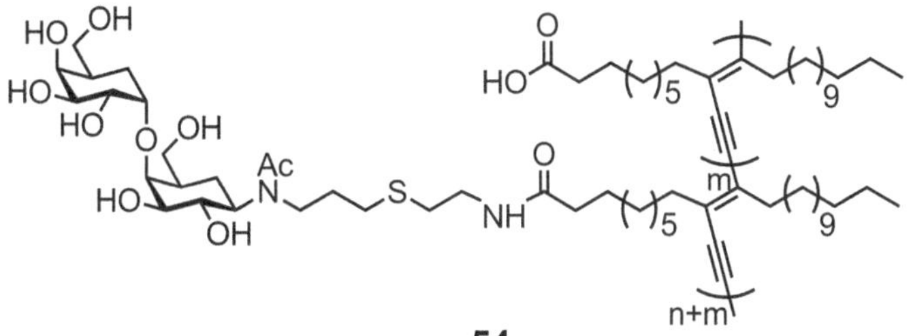

**54**

**Fig. 2.13** Chemical structure of CPE 54

developed a solid-phase chip based on streptavidin-functionalized PDA liposome for the diagnosis of pathogen infection [52]. The method takes advantage of the unique optical property of PDA and the specific biotin-STA interactions to detect highly infectious *C. trachomatis*. The biotinylated nucleic acids, obtained from the PCR amplification of a *C. trachomatis*-specific sequence by introducing biotin-11-dUTP, could successfully accomplish the specific transformation of PDA from nonfluorescent blue state (633 nm) to fluorescent red state (547 nm). In addition to the specific recognition of pathogens, other methods without access to the detailed biochemical information and structure characteristics of microorganisms have also been developed [53].

Wang and co-workers realized the visual optical discrimination and detection of microbial pathogens by using a blend of two cationic CPEs [54]. Cationic CPEs PFP 5 and PPV 55 (chemical structure shown in Fig. 2.14) were involved, and FRET between PFP 5 and PPV 55 could be induced after the addition of multi-negatively charged species. The chosen representative microorganisms were *C. albicans* fungus and *E. coli* bacterium. Driven by electrostatic interactions, the two cationic CPEs were bound to the negatively charged microbial membrane surfaces, which made them close enough to favor microorganism-mediated FRET. Because of the different surface structures and chemical compositions of the cell

**Fig. 2.14** Chemical structure of PPV 55

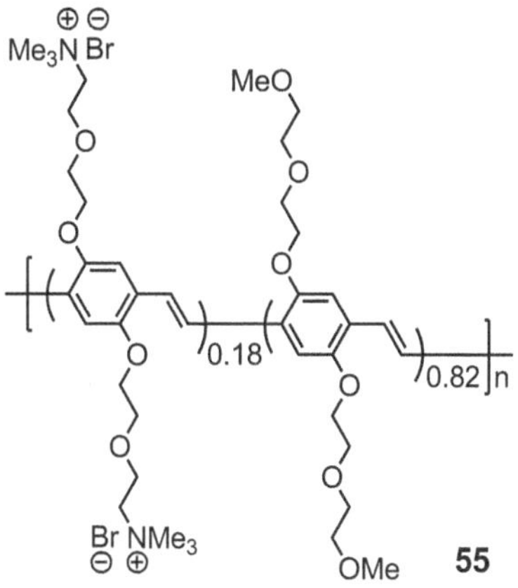

**55**

walls between fungi and bacteria, two polymers exhibited different association situations. Compared to *E. coli*, *C. albicans* captured more PFP 5 and less PPV 55. Thus, the imbalanced binding led to distinct FRET, weak FRET for *C. albicans* and strong FRET for *E. coli*. Because PPV 55 was designed to bear OEG side-chains so as to resist nonspecific adsorption to proteins and cell surfaces, PPV 55 possessed greater possibility to be responsible for the differential interactions between fungi and bacteria. In addition, combined with PCR amplification of a specific region of *C. albicans*, this method could be used for specific detection of clinically important *C. albicans* by DNA-mediated FRET, which was independent of electrophoresis analyses and real-time PCR.

Chemical nose/tongue strategies were feasible for the detection and discrimination of bacteria. Bunz and co-workers have initiated a rapid method to simultaneously detect and discriminate multiple bacteria with the aid of gold NPs-PPE complexes [55]. The anionic PPE 31 could form electrostatic complexes with cationic gold NPs, in which the fluorescence of PPE 31 was efficiently quenched by gold NPs through an electron/energy-transfer process. Upon adding negatively charged bacteria to nonfluorescent gold NPs-PPE complexes, the anionic PPE 31 was competitively displaced from gold surfaces and recovered its initial fluorescence. PPE 31 was suitable to sense bacteria because of its superior ability to provide multivalent interaction sites and excellent molecular wire effect. Ammonium-functionalized gold NPs (NP1-NP3) with different hydrophobic abilities were employed to regulate diverse interactions with the hydrophobic or functional microbial surfaces. After constructing the quenched electrostatic complexes, 12 bacteria were added into the system to mediate the characteristic fluorescence changes. In combination with LDA, the raw fluorescence responses (the net difference of fluorescence intensity, $\Delta I$) could be converted into several canonical scores that are the linear combinations of response patterns and mainly responsible for the discrimination and assignment of bacteria even with tiny differences.

**Fig. 2.15** Canonical score plot for the fluorescence response patterns as determined with LDA (Reprinted with permission from Ref. [55]. Copyright (2008) Wiely-VCH Verlag GmBH & Co. KGaA Weinheim)

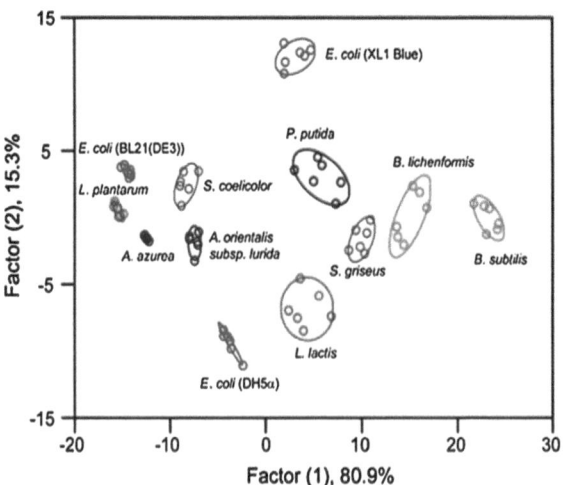

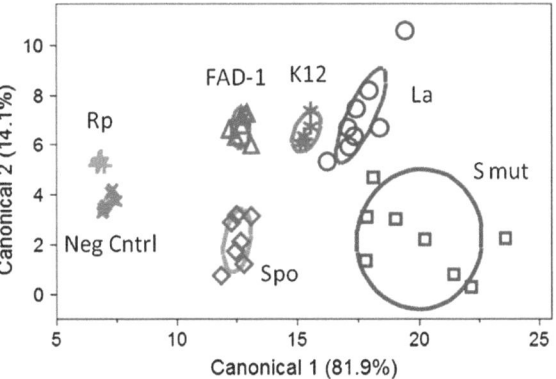

**Fig. 2.16** Canonical score plot of the response pattern for the discrimination of various pathogens (Reprinted with permission from Ref. [56]. Copyright (2010) American Chemical Society)

In this case, all twelve bacteria could be completely differentiated to be located in the specific areas of the discriminant graph as shown in Fig. 2.15. In buffered solution, the test was effective not only for Gram-positive bacteria (e.g., *A. azurea* and *B. subtilis*) but also for Gram-negative bacteria (e.g., *E. coli* and *P. putida*). Even for different strains of the identical bacteria (*E. coli* BL21(DE), DH5α, and XL1 Blue), the method still worked very well. By comparing the distances of test specimens to the centroids of the obtained training matrix (i.e., shortest Mahalanobis distance), 64 unknown samples randomly selected from the 12 bacteria could be successfully assigned with a 95 % accuracy.

Bazan and co-workers developed a sensitive FRET-based method to recognize various bacteria by using cationic conjugated oligoelectrolyte 37/Fl-labeled single-stranded DNA (ssDNA) electrostatic complexes [56]. Additions of gram-negative (*Escherichia coli* K12, *Escherichia coli* FAD-1, *Sporomusa* DMG58) and gram-positive (*Lactobacillus acidophilus*, *Rhodopseudomonas palustris* CGA009, *Streptococcus mutans*) bacteria led to modifications of the FRET efficiency that were dependent on the ssDNA sequence. Transformation of the acquired response patterns by regularized discriminant analysis (RDA) could divide the bacteria into several classes without any misclassification according to the first two canonical scores (Fig. 2.16).

## 2.3  Diagnostics of Tumor

Rapid, sensitive and selective detection of cancer cells is a great challenge for the early diagnosis and treatment of tumors [57]. Current methods employed for cancer cell detection are in general based on DNA microarray and antibody array techniques, which rely on mutations in the genome and variations in intra and extracellular protein biomarkers, respectively. However, they are not specific enough to distinguish normal and cancer cells. Highly sensitive diagnosis of tumor cells with CPEs is an appealing research area. In 2009, Bunz and co-workers utilized the concept of chemical nose/tongue to implement the rapid and sensitive detection and discrimination of normal and cancer cells by gold NP-PPE sensors, in which the unique molecular signatures are not required [58]. The basic principle was similar to the previous study for bacteria discrimination. Cationic gold NPs and anionic PPE 32 formed nonfluorescent electrostatic complexes. The negatively charged normal or cancer cells with subtle surface differences (generally varying compositions of phospholipids, carbohydrates, and proteins) influenced the association of gold NPs and PPEs via complicated interactions (e.g., charge and hydrophobicity) and displaced PPE 32 from the gold NP surface to differentially recover its fluorescence. By virtue of LDA, the diverse response patterns (quantified as fluorescence intensity change) collected by three gold NP-PPE sensors were converted into canonical fluorescence-response plots. In the discriminant plots, cells were clustered into several independent groups according to the different cell types, including (i) different cancer cells; (ii) normal, cancerous, and metastatic human breast cells; and (iii) isogenic normal, cancerous, and metastatic murine epithelial cells. Therefore, the objective of discrimination among different cell types was successfully accomplished. Given that the 95 % confidence ellipses of normal cells (MCF-10A and CDBgeo) and metastatic cells (V14 and MDA-MB-231) were independently overlapping, a potential correlation between different diseases states was demonstrated (Fig. 2.17).

By replacing gold NPs-PPE with eight functionalized PPEs, Bunz and co-workers successfully raised another approach toward cell discrimination between normal and cancer cells [59]. The eight functionalized PPEs possessed different charge characteristics, polymerization degrees, and hydrophobicity. The differential interactions between cell membranes and cationic/anionic PPEs generated diverse response patterns quantified as relative fluorescence changes ($I/I_0$) of polymers. After optimizing the polymer combinations with the highest

**Fig. 2.17** Canonical score plot for the first two factors of fluorescence-response pattern against different normal and cancerous cell types (Reprinted with permission from Ref. [58]. Copyright (2009) National Academy of Sciences, USA)

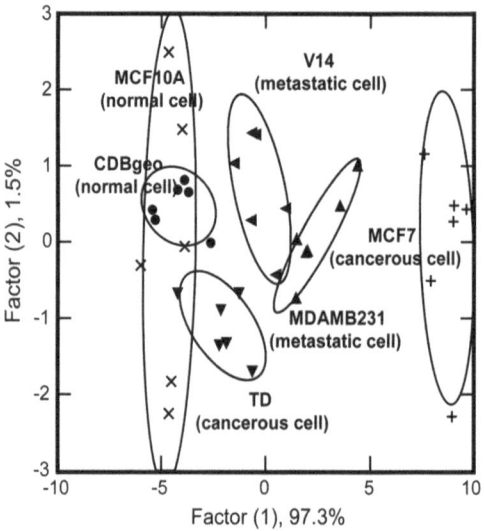

differentiation accuracy, the number of sensing elements could be reduced from eight to four. By combining with LDA, different cell types, including four human cancer cells, three normal, cancerous, and metastatic isogenic mouse cells could be easily identified. Moreover, according to the shortest Mahalonobis distance to the pre-established discrimination plot, an 80 % accuracy of unknown samples from the three isogenic cell lines was shown. As a further development to tumor cell detection, malignant and metastatic tumor cells sorting seem more attractive in terms of the evaluation of prognosis and antitumor treatment effects.

Recently, based on a multifunctional NP composite system, Feng and co-workers developed a new method for the detection and sorting of circulating tumor cells (CTC) with over-expressed cell-surface receptors HER2 [60]. The employed magnetic fluorescent composite NPs (MF-NPs, 100 nm in diameter) contained four independent but closely relevant elements: (i) conjugated oligomer-terminated POSS (COPOSS, nonquaternized form of 56, chemical structure shown in Fig. 2.18) as the high-performance fluorescent probe, (ii) silica NPs as a formulation matrix for mounting COPOSS and avoiding the unexpected fluorescence quenching from iron oxide, (iii) iron oxide layer around silica NPs providing a function for magnetic-guided sorting, and (iv) outermost herceptin as the specific antibody for targeting recognition of HER2. On the basis of the differential expression of HER2 between SK-BR-3 (high) and NIH 3T3 (low) cells, the low cytotoxic MF-NPs were observed to be considerably internalized by SK-BR-3 cells over NIH 3T3 cells via RME. Further, under an external magnetic field, SK-BR-3 cells accumulated with a large amount of magnetic particles could be selectively separated and collected from a mixture of SK-BR-3 and NIH 3T3 cells.

**Fig. 2.18** Chemical structure of COPOSS 56

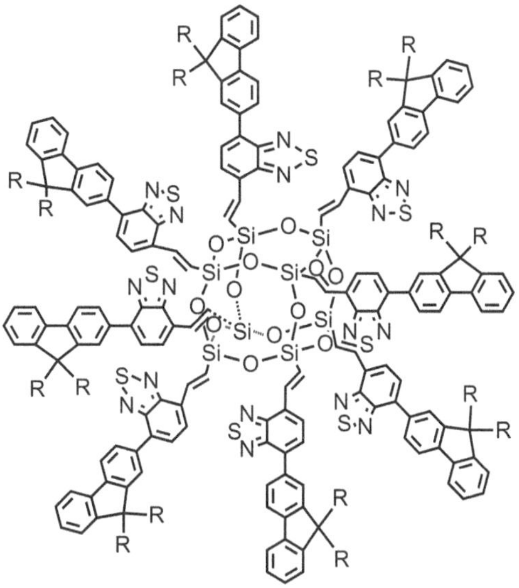

$R = (CH_2)_6NMe_3Br$

**56**

Because the composite NPs possessed both magnetic and fluorescent properties, the process of cell detection and real-time cell sorting could be simultaneously finished.

The well-developed FRET between PFP 5 with Fl-labeled DNA extended to be employed for cancer cell detection. Wang and co-workers developed a target cell-specific aptamers PCR approach for sensitive and selective detection of acute myeloid leukemia (HL-60) cells [61]. One advantage of this strategy is that DNA aptamers function as recognition elements for target cells and, meanwhile, as the template for PCR amplification, which means that the detection protocols are easy to handle by eliminating the need for intensive sample preparation, labeling, and separation processes. The other advantage of the approach is the combination of signal amplification of conjugated polymers and PCR amplification to achieve very good detection sensitivity with a single cell detection limit.

# References

1. Dubertret B, Calame M, Libchaber AJ (2001) Single-mismatch detection using gold-quenched fluorescent oligonucleotides. Nat Biotech 19:365–370
2. Wang J (2000) Survey and summary. Nucleic Acids Res 28:3011–3016
3. Daar AS, Thorsteinsdottir H, Martin DK, Smith AC, Nast S, Singer PA (2002) Top ten bio-technologies for improving health in developing countries. Nat Genet 32:229–232
4. Expression profiling-best practices for data generation and interpretation in clinical trials. (2004) Nat Rev Genet 5:229–237
5. Gaylord BS, Heeger AJ, Bazan GC (2002) DNA detection using water-soluble conjugated polymers and peptide nucleic acid probes. Proc Natl Acad Sci USA 99:10954–10957
6. Gaylord BS, Massie MR, Feinstein SC, Bazan GC (2005) SNP detection using peptide nucleic acid probes and conjugated polymers: applications in neurodegenerative disease identification. Proc Natl Acad Sci USA 102:34–39
7. Al Attar HA, Norden J, O'Brien S, Monkman AP (2008) Improved single nucleotide poly-morphisms detection using conjugated polymer/surfactant system and peptide nucleic acid. Biosens Bioelectron 23:1466–1472
8. Kim S, Misra A (2007) SNP genotyping: technologies and biomedical applications. Annu Rev Biomed Eng 9:289–320
9. Ding C (2007) Other applications of single nucleotide polymorphisms. Trends Biotechnol 25:279–283
10. The International HapMap C (2005) A haplotype map of the human genome. Nature 437:1299–1320
11. Li K, Liu B (2009) Conjugated polyelectrolyte amplified thiazole orange emission for label free sequence specific DNA detection with single nucleotide polymorphism selectivity. Anal Chem 81:4099–4105
12. Duan X, Li Z, He F, Wang S (2007) A sensitive and homogeneous SNP detection using cationic conjugated polymers. J Am Chem Soc 129:4154–4155
13. Duan X, Yue W, Liu L, Li Z, Li Y, He F, Zhu D, Zhou G, Wang S (2009) Single-nucleotide polymorphism (SNP) genotyping using cationic conjugated polymers in homogeneous solution. Nat Protocols 4:984–991
14. Duan X, Liu L, Wang S (2009) Homogeneous and one-step fluorescent allele-specific PCR for SNP genotyping assays using conjugated polyelectrolytes. Biosens Bioelectron 24:2095–2099
15. Duan X, Wang S, Li Z (2008) Conjugated polyelectrolyte-DNA complexes for multi-color and one-tube SNP genotyping assays. Chem Commun 44:1302–1304

16. Najari A, Ho HA, Gravel J-F, Nobert P, Boudreau D, Leclerc M (2006) Reagentless ultra-sensitive specific DNA array detection based on responsive polymeric biochips. Anal Chem 78:7896–7899
17. Wang C, Zhan R, Pu K-Y, Liu B (2010) Cationic polyelectrolyte amplified bead array for DNA detection with zeptomole sensitivity and single nucleotide polymorphism selectivity. Adv Funct Mater 20:2597–2604
18. Bird A (2002) DNA methylation patterns and epigenetic memory. Genes Dev 16:6–21
19. Reik W, Walter J (2001) Genomic imprinting: parental influence on the genome. Nat Rev Genet 2:21–32
20. Robertson KD, Wolffe AP (2000) DNA methylation in health and disease. Nat Rev Genet 1:11–19
21. Feng F, Wang H, Han L, Wang S (2008) Fluorescent conjugated polyelectrolyte as an indicator for convenient detection of DNA methylation. J Am Chem Soc 130:11338–11343
22. Sadri R, Hornsby PJ (1996) Rapid analysis of DNA methylation using new restriction enzyme sites created by bisulfite modification. Nucleic Acids Res 24:5058–5059
23. Feng F, Liu L, Wang S (2010) Fluorescent conjugated polymer-based FRET technique for detection of DNA methylation of cancer cells. Nat Protocols 5:1255–1264
24. Yang Q, Dong Y, Wu W, Zhu C, Chong H, Lu J, Yu D, Liu L, Lv F, Wang S (2012) Detection and differential diagnosis of colon cancer by a cumulative analysis of promoter methylation. Nat Commun 3:1206
25. Yang Q, Qiu T, Wu W, Zhu C, Liu L, Ying J, Wang S (2011) Simple and sensitive method for detecting point mutations of epidermal growth factor receptor using cationic conjugated polymers. ACS Appl Mater Interfaces 3:4539–4545
26. Song J, Yang Q, Lv F, Liu L, Wang S (2012) Visual detection of DNA mutation using multi-color fluorescent coding. ACS Appl Mater Interfaces 4:2885–2890
27. Aneja A, Mathur N, Bhatnagar PK, Mathur PC (2009) Detection of known mutations for medical diagnostics by FRET spectroscopy. J Biomater Sci Polym Ed 20:1823–1830
28. Guan H, Zhou P, Zeng S, Zhou X, Wang Y, He Z (2009) Detection of deletion mutations in DNA using water-soluble cationic fluorescent thiophene copolymer. Talanta 79:153–158
29. Feng F, Duan X, Wang S (2009) Fluorescence-amplifying assay for irradiated DNA lesions using water-soluble conjugated polymers. Macromol Rapid Commun 30:147–151
30. Kwon IK, Kim JP, Sim SJ (2010) Enhancement of sensitivity using hybrid stimulus for the diagnosis of prostate cancer based on polydiacetylene (PDA) supramolecules. Biosens Bioelectron 26:1548–1553
31. Zheng W, He L (2010) Multiplexed detection of protein cancer markers on au/ag-barcoded nanorods using fluorescent-conjugated polymers. Anal Bioanal Chem 397:2261–2270
32. Yao Z, Feng X, Li C, Shi G (2009) Conjugated polyelectrolyte as a colorimetric and fluorescent probe for the detection of glutathione. Chem Commun 45:5886–5888
33. Fan H, Zhang T, Lv S, Jin Q (2010) Fluorescence turn-on assay for glutathione reductase activity based on a conjugated polyelectrolyte with multiple carboxylate groups. J Mater Chem 20:10901–10907
34. Wang Y, Liu B (2009) Conjugated polyelectrolyte-sensitized fluorescent detection of thrombin in blood serum using aptamer-immobilized silica nanoparticles as the platform. Langmuir 25:12787–12793
35. You C–C, Miranda OR, Gider B, Ghosh PS, Kim I-B, Erdogan B, Krovi SA, Bunz UHF, Rotello VM (2007) Detection and identification of proteins using nanoparticle-fluorescent polymer 'chemical nose' sensors. Nat Nanotech 2:318–323
36. Miranda OR, You C–C, Phillips R, Kim I-B, Ghosh PS, Bunz UHF, Rotello VM (2007) Array-based sensing of proteins using conjugated polymers. J Am Chem Soc 129:9856–9857
37. Li H, Bazan GC (2009) Conjugated oligoelectrolyte/ssdna aggregates: self-assembled multi-component chromophores for protein discrimination. Adv Mater 21:964–967
38. Niamnont N, Mungkarndee R, Techakriengkrai I, Rashatasakhon P, Sukwattanasinitt M (2010) Protein discrimination by fluorescent sensor array constituted of variously charged dendritic phenylene–ethynylene fluorophores. Biosens Bioelectron 26:863–867

39. Xu Q, Wu C, Zhu C, Duan X, Liu L, Han Y, Wang Y, Wang S (2010) A water-soluble con-jugated polymer for protein identification and denaturation detection. Chem Asian J 5:2524–2529
40. Nilsson KPR, Herland A, Hammarstrom P, Inganas O (2005) Conjugated polyelectrolytes: conformation-sensitive optical probes for detection of arnyloid fibril forrnation. Biochemistry 44:3718–3724
41. Herland A, Nilsson KPR, Olsson JDM, Hammarström P, Konradsson P, Inganäs O (2005) Synthesis of a regioregular zwitterionic conjugated oligoelectrolyte, usable as an optical probe for detection of amyloid fibril formation at acidic pH. J Am Chem Soc 127:2317–2323
42. Wigenius J, Andersson MR, Esbjoerner EK, Westerlund F (2011) Interactions between a luminescent conjugated polyelectrolyte and amyloid fibrils investigated with flow linear dichroism spectroscopy. Biochem Biophys Res Commun 408:115–119
43. Stabo-Eeg F, Lindgren M, Nilsson KPR, Inganas O, Hammarstrom P (2007) Quantum effi-ciency and two-photon absorption cross-section of conjugated polyelectrolytes used for protein conformation measurements with applications on amyloid structures. Chem Phys 336:121–126
44. Wigenius J, Persson G, Widengren J, Inganas O (2011) Interactions between a luminescent conjugated oligoelectrolyte and insulin during early phases of amyloid formation. Macromol Biosci 11:1120–1127
45. Disney MD, Zheng J, Swager TM, Seeberger PH (2004) Detection of bacteria with carbohy-drate-functionalized fluorescent polymers. J Am Chem Soc 126:13343–13346
46. Baek M-G, Stevens RC, Charych DH (2000) Design and synthesis of novel glycopolythio-phene assemblies for colorimetric detection of influenza virus and E. Coli. Bioconjugate Chem 11:777–788
47. Xue C, Velayudham S, Johnson S, Saha R, Smith A, Brewer W, Murthy P, Bagley ST, Liu H (2009) Highly water-soluble, fluorescent, conjugated fluorene-based glycopolymers with poly(ethylene glycol)-tethered spacers for sensitive detection of E coli. Chem Eur J 15:2289–2295
48. Ho H-A, Boissinot M, Bergeron MG, Corbeil G, Doré K, Boudreau D, Leclerc M (2002) Colorimetric and fluorometric detection of nucleic acids using cationic polythiophene deriva-tives. Angew Chem Int Ed 41:1548–1551
49. Guan H, Cai M, Chen L, Wang Y, He Z (2010) Label-free DNA sensor based on fluores-cent cationic polythiophene for the sensitive detection of hepatitis B virus oligonucleotides. Luminescence 25:311–316
50. Lee SW, Kang CD, Yang DH, Lee J-S, Kim JM, Ahn DJ, Sim SJ (2007) The development of a generic bioanalytical matrix using polydiacetylenes. Adv Funct Mater 17:2038–2044
51. Beaulieu C, Guay D, Wang Z, Leblanc Y, Roy P, Dufresne C, Zamboni R, Berthelette C, Day S, Tsou N, Denis D, Greig G, Mathieu M-C, O'Neill G (2008) Identification of prostaglandin D-2 receptor antagonists based on a tetrahydropyridoindole scaffold. Bioorg Med Chem Lett 18:2696–2700
52. Jung YK, Kim TW, Jung C, Cho D-Y, Park HG (2008) A polydiacetylene microchip based on a biotin-streptavidin interaction for the diagnosis of pathogen infections. Small 4:1778–1784
53. Meir D, Silbert L, Volinsky R, Kolusheva S, Weiser I, Jelinek R (2008) Colorimetric/fluo-rescent bacterial sensing by agarose-embedded lipid/polydiacetylene films. J Appl Microbiol 104:787–795
54. Zhu C, Yang Q, Liu L, Wang S (2011) Visual optical discrimination and detection of micro-bial pathogens based on diverse interactions of conjugated polyelectrolytes with cells. J Mater Chem 21:7905–7912
55. Phillips RL, Miranda OR, You C-C, Rotello VM, Bunz UHF (2008) Rapid and efficient iden-tification of bacteria using gold-nanoparticle–poly(para-phenyleneethynylene) constructs. Angew Chem Int Ed 47:2590–2594
56. Duarte A, Chworos A, Flagan SF, Hanrahan G, Bazan GC (2010) Identification of bacteria by conjugated oligoelectrolyte/single-stranded DNA electrostatic complexes. J Am Chem Soc 132:12562–12564

57. Wulfkuhle JD, Liotta LA, Petricoin EF (2003) Proteomic applications for the early detection of cancer. Nat Rev Cancer 3:267–275

58. Bajaj A, Miranda OR, Kim IB, Phillips RL, Jerry DJ, Bunz UHF, Rotello VM (2009) Detection and differentiation of normal, cancerous, and metastatic cells using nanoparticle-polymer sensor arrays. Proc Natl Acad Sci USA 106:10912–10916

59. Bajaj A, Miranda OR, Phillips R, Kim I-B, Jerry DJ, Bunz UHF, Rotello VM (2009) Array-based sensing of normal, cancerous, and metastatic cells using conjugated fluorescent polymers. J Am Chem Soc 132:1018–1022

60. Mi Y, Li K, Liu Y, Pu K-Y, Liu B, Feng S-S (2011) Herceptin functionalized polyhedral oligomeric silsesquioxane-conjugated oligomers-silica/iron oxide nanoparticles for tumor cell sorting and detection. Biomaterials 32:8226–8233

61. Song J, Lv F, Yang G, Liu L, Yang Q, Wang S (2012) Aptamer-based polymerase chain reaction for ultrasensitive cell detection. Chem Commun 48:7465–7467

# Chapter 3
# Fluorescence Imaging Applications of Functionalized Conjugated Polyelectrolytes

**Abstract** CPEs have established themselves as useful imaging agents for live cells due to their superior properties of large extinction coefficients, good photo-stability, large two-photon action cross sections, low cytotoxicity, and versatile surface modification. In this chapter, fluorescence imaging applications of func-tionalized conjugated polyelectrolytes are presented, where specific cells imaging have been successfully achieved by functionalizing CPEs with specific recogni-tion elements to attain selective binding. Specific cellular organelle imaging has also been successfully achieved with CPEs. As a new water-dispersible form of conjugated polymers, conjugated polymer nanoparticles present great potential in biological imaging due to their superior properties, which are comprehensively introduced in this chapter.

**Keywords** Fluorescence imaging • Conjugated polymer nanoparticles • Target cell imaging • In vitro imaging • Ex/in vivo imaging

Imaging techniques are a vital part of clinical diagnostics, biomedical research, and nanotechnology. Over the past decade, fluorescence molecular imaging has been increasingly used to diagnose diseased tissue, to visualize dynamic thera-peutic effects, and to monitor therapeutic outcomes using specific biological and molecular targets [1]. A number of in vivo imaging modalities and molecular tar-geted contrast agents are coming of age and provide a versatile platform on which to design and implement molecular imaging. For cellular imaging, it is important not only to develop a safe and bright fluorescent probes but also to functional-ize them with appropriate targeting ligand to image specific cells. Imaging with CPEs in vitro could be classified into two types, namely, nonspecific recognition and specific recognition. Nonspecific recognition means that no specific target-ing elements are linked to polymers but directly used for imaging. The involved mechanism is nonselective cellular uptake. Bunz and co-workers reported ani-onic PPE 35 for imaging fibronectin fibrils of live NIH 3T3 fibroblast cells [2].

Despite that no specific ligands were linked to PPE 35, selective binding toward extracellular matrix (ECM) protein fibronectin was still achieved after 4 h incubation, which was confirmed by fluorescent colocalization with antifibronectin antibodies. According to the calculated apparent dissociation constant (100 nm) of fibronectin/PPE 35 complex, a low-affinity polyvalent recognition mechanism was proposed. Because there are several positively charged sites in fibronectin to accommodate polyanionic heparin, the competitive binding of heparin to PPE 35 was shown. Moreover, 2P imaging of fibronectin with 35 was also demonstrated. Extending incubation time could lead to the internalization of anionic 35 as well, which is typically difficult to occur in live cell systems. Taking cationic PPE 33 as a contrast, a distinct internalization pattern was observed for both 4 and 24 h, indicating the specific binding of polymer 35 to fibronectin.

OEG-pendent PPV 3 was specially designed for the detection and imaging of apoptotic cells [3]. Although no specific groups were linked to polymer 3, OEG side-chains and QA groups featured PPV 3 still exhibited exclusive binding to apoptotic cells. By inducing Jurkat T cells to suffer from apoptosis with anti-Fas mAbs, a series of apoptosis-related changes in the cells occurred. Taking advantage of the increased negative charge density and enhanced membrane permeability in apoptotic cells, the low cytotoxic PPV 3 was efficient to selectively bind and successively enter into the cells. In contrast, normal cells without treatment with anti-Fas mAbs were seldom stained by PPV 3 as a result of the effect of OEG side-chains to eliminate nonspecific interactions.

Conjugated polymer nanoparticles (CPNs) are water-dispersible form of conjugated polymers, which have aroused great attentions for imaging applications [4]. It has reported that CPNs have several unique advantages, such as high brightness, large extinction coefficients, superior photostability, large 2P action cross sections, low cytotoxicity, facile chemical synthesis, tunable spectral properties, and versatile surface modification. In addition, both size-dependent photophysical properties and fluorescence blinking phenomenon are not observed for CPNs, which is superior to quantum dots (QDs).

Moon et al. prepared stable PPE-based CPNs via phase inversion precipitation of polymer 4 in a poor solvent for live cell imaging [5]. The obtained CPNs exhibited an average size of 93 nm with a QY of 0.17 in water. Both live and fixed BALB/C 3T3 cells were successfully stained by CPNs. The location was mainly in the cytosol, especially around the perinuclear region. Cell viability analysis indicated that CPNs basically showed no cytotoxicity to baby hamster kidney (BHK) cells over 1 week of culture. Photobleaching experiment in fixed BALB/C 3T3 cells demonstrated that CPNs possessed excellent photostability, which was ascribed to the assumption that the hydrophobic side-chains of polymer 4 constituted a protective layer that prevented the destructive ROS from diffusing into the internal conjugated backbones. Utilizing PPE 4 CPNs, 2P imaging of endothelial cells in a tissue model was realized on the basis of the exceptional nonlinear optical property of CPNs [6].

PEG-capped poly[2-(2′,5′-bis-(2″-ethylhexyloxy)phenyl)-1,4-phenylene vinylene] (BEHPPPV) nanospheres and surfactant-capped MEH-PPV CPNs were prepared by Green and co-workers for intracellular imaging. To further improve the biocompability and provide active sites for subsequent biological labeling, they reported PEG-phospholipid encapsulated CPNs 1,2-diacyl-sn-glycero-3-phos-phoethanolamine-N-[methoxy(polyethylene glycol)-2000] (PEG2000-PE)/1,2-dipalmitoyl-sn-glycero-3-phosphocholine (DPPC) (PEG2000-PE/DPPC micelles) for cell imaging and protein attachment [7]. In addition, the conjugation of carboxylate-functionalized CPNs with BSA via carbodiimide (EDC)/sulfo-NHS reaction was perfectly realized, providing the basis for targeting imaging.

The rigid PF internal core and the flexible PEG external shell made hyperbranched CPE (HCPE, 19) intrinsically form single molecular core-shell nanospheres (10.7 nm in diameter) with a quantum yield of 30 % in buffer solution. Unimolecular cationic oligofluorene-substituted polyhedral oligomeric silsesquioxane (POSS) NPs were fabricated [8]. Cationic oligofluorene and POSS were linked together via Heck coupling, affording the water-soluble single-molecular NPs (cationic oligofluorene substituted POSS (OFP)) with an average diameter of 3.6 nm and a quantum yield of 0.85 in water. The signal amplification effect provided by OFP through FRET was investigated in fixed MCF-7 cells. Interestingly, after incubating cells with OPF for 2 h, blue fluorescence was found in the entire cells including the nucleus which was attributed to its small size to pass through the nuclear pore complex.

Uniformly dispersed CPNs were developed by Li and co-workers [9], CPEs were coated on well-defined core-shell Ag@SiO$_2$ NPs via electrostatic self-assembly technique and the fabricated CPEs NPs were utilized for cell imaging. Similarly, Schanze and co-workers reported the fabrication of PPE-coated silica NPs by layer-by-layer technique for 2P cellular imaging [10].

With the continuous development of CPNs for cellular fluorescent imaging, multicolor and multifunctionality cell imaging have been delicately introduced. McNeill and co-workers reported the multicolor fluorescence imaging in living cells with five different colored CPNs (poly(9,9-dioctylfluorenyl-2,7-diyl) (PFO), PPE 86, poly[{9,9-dioctyl-2,7-divinylene-fluorenylene}-alt-co-{2-methoxy-5-(2-ethylhexyloxy)-1,4-phenylene}]] (PFPV) (poly[(9,9-dioctylfluorenyl-2,7-diyl)-co-(1,4-benzo-{2,1′,3}-thiadiazole)] (PFBT), and poly[2-methoxy-5-(2-ethylhexyloxy)-1,4-phenylenevinylene] (MEH-PPV)) prepared by the reprecipitation method [11]. Compared to QDs with the same size (15 nm diameter), the obtained multicolor CPNs showed markedly improved fluorescence brightness, radiative rates, and photostability, decreased or no blinking, and ultrahigh 2P action cross section. The multicolor CPNs internalized by living J774.A1 macrophage cells were mainly located in the cytoplasm. The cellular uptake mechanism of highly fluorescent CPNs was further investigated and verified that macropinocytosis was the primary endocytic mechanism. Owing to the small particle size, large optical cross section, bright phosphorescence, high oxygen sensitivity, good pH stability, low cytotoxicity, and good cellular uptake properties, oxygen-sensitive phosphorescent dye (PtOEP)-doped CPNs was successfully utilizes to sense oxygen in living cells [12].

Specific cell imaging, or targeting imaging, is to modify CPEs with specific recognition elements (i.e., antibody, peptide, sugar, protein, etc.) to attain selective binding, which is becoming the current research focus. Bunz and co-workers realized specific cell imaging by using a designed anionic folate-functionalized PPE 7 [13]. In contrast to FR-negative NIH 3T3 cells, the obtained polymer was able to selectively recognize and image FR-overexpressed KB cells.

Aforementioned water-soluble hyperbranched CPE conjugated with anti-HER2 affibody was synthesized and successfully realized targeted cell imaging toward HER2 over-expressed SK-BR-3 cells [14]. Chiu and co-workers modified the carboxyl-functionalized ultrabright PFBT CPNs with STA and IgG to perform specific cell imaging [15].

To realize targeted nucleus imaging of cancer cells, Liu and co-workers constructed pH-responsive CPNs that were composed of fluorescent probe 56, pH sensitive biopolymer chitosan, and specific ligand folate [16]. After specifically being internalized by FR-positive MCF-7 cells via folate receptor-mediated endocytosis, the pH responsive CPNs were subjected to rapid release of COE POSS in acidic compartments (pH ≈ 5.0, late endosomes and/or lysosomes), which initiated the following targeted nucleus staining. They introduced biocompatible poly(DL-lactide-co-glycolide) (PLGA) to functionalize CPNs, yielding four different CPNs with emission colors from blue to red [17]. Compared to pure CPNs, folate-functionalized CPNs showed an enhanced internalization by FR-overexpressed MCF-7 cells. Analogously, they also showed that anti-HER2 mAb-functionalized CPNs could be used for HER2-positive cancer cell imaging by virtue of the specific antibody-antigen recognition [18]. To develop a simple multiplex imaging strategy, Liu and co-workers designed multicolor RGD (Arg–Gly–Asp)-functionalized CPNs for targeting cell imaging [19]. Furthrmore, they utilized surface-functionalized CPNs to achieve the discrimination and imaging of mixed living cancer cells in a single solution [20].

Barbarella and co-workers prepared a series of oligo-PTs (57–64, chemical structure shown in Fig. 3.1) with different moieties including 4-sulfo-2,3,5,6-tetrafluorophenyl

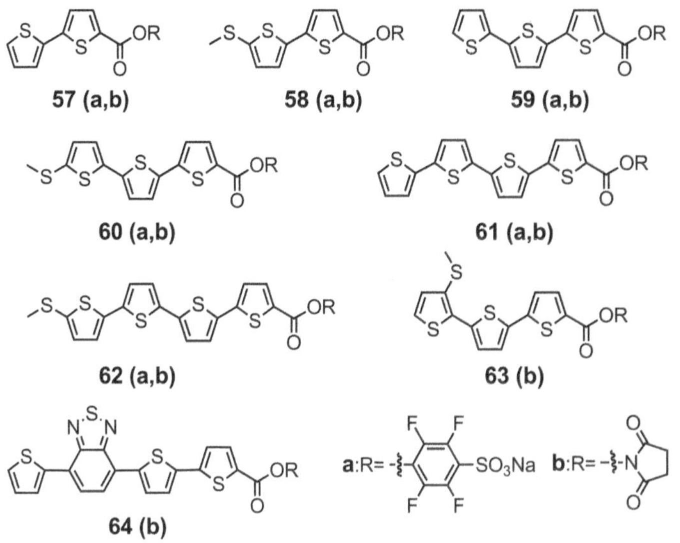

**Fig. 3.1** Chemical structures of CPEs 57–64

esters (PT-STP, a) and N-hydroxysuccinimidyl esters (PT-NHS, b) [21]. Their photophysical properties were almost the same and all of them exhibited superior photostability. Through the reactions between anti-CD38 or anti-CD4 mAbs and PT-STP or PTNHS (emission colors from blue to red), different fluorescent labeled anti-CD38 or anti-CD4 mAbs were obtained. Targeting imaging toward fixed MOLT 4 cells (expressing glycoprotein CD-4 antigen) was successfully realized.

The functionalization of PFBT-CPNs was realized by incorporating PEG lipids with different functional end groups and developed for extracellular targeting

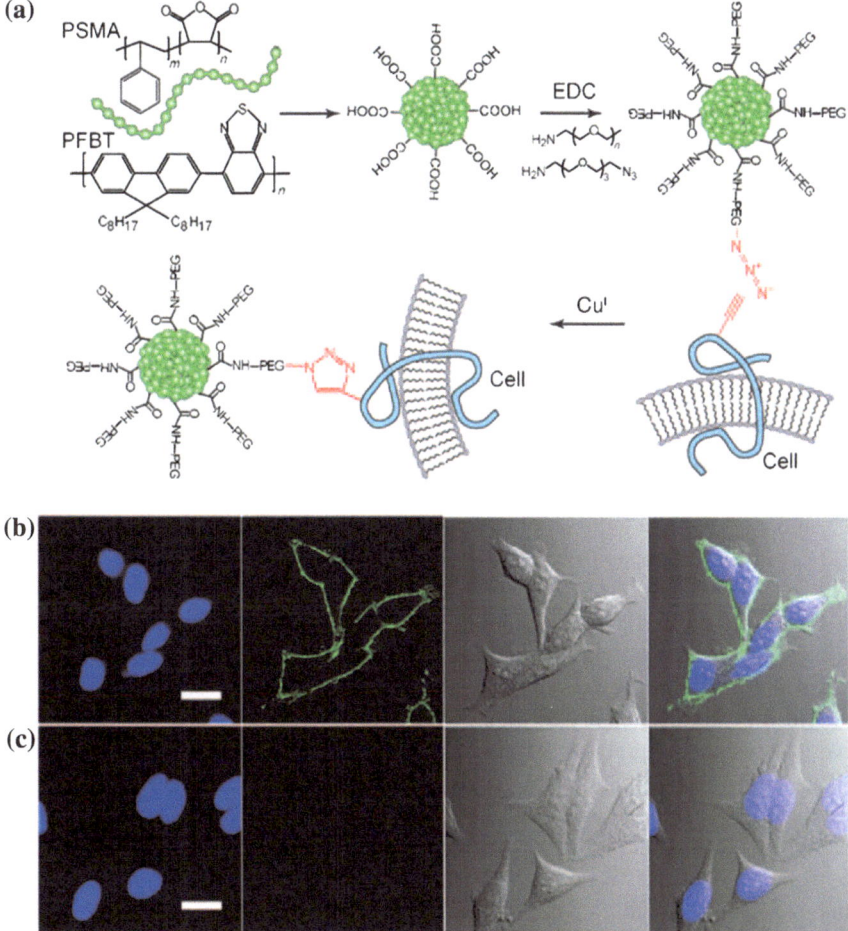

Fig. 3.2 **a** Schematic representation of the fabrication and functionalization of CPNs for bioorthogonal labeling cell surface using click chemistry. **b** CLSM images of CPN-alkyne-labeled AHA-treated MCF-7 cells in the presence of reducing agents (sodium ascorbate); **c** in the absence of reducing agents. The *left four panels* show fluorescence images; *green* fluorescence is from PFBT-CPNs, and *blue* fluorescence is from the nuclear stain Hoechst 34580. The *right four panels* show DIC and combined DIC/fluorescence images. Scale bars: 20 μm (Reprinted with permission from Ref. [23]. Copyright (2010) Wiely-VCH Verlag GmBH & Co. KGaA Weinheim)

labeling [22]. With biotinylated anti-CD16/32 mAb as a primary binding element and STA as a sandwich bridge molecule, the specific imaging of surface receptor CD16/32 of J774A.1 cells was perfectly achieved. To facilitate the surface modification of CPNs, Chiu and co-workers used a maleic anhydride-containing amphiphilic polymer poly(styrene-co-maleic anhydride) (PSMA) to functionalize PFBT CPNs with carboxyl groups for further conjugation with biological molecules (Fig. 3.2a) [23]. The click reaction was utilized to label specific cellular targets in a highly selective manner. The resultant CPNs exhibited high brightness, excellent photostability, and resistance to fluorescence bleaching. By bioorthogonal noncanonical amino acid-tagging (BONCAT) technique, the proteins in the cells could be modified with alkyne- or azido bearing artificial amino acids (AHA and HPG). As is seen in CLSM images (Fig. 3.2b, c), bioorthogonal labeling with alkyne-CPNs mediated by click reaction was successfully achieved in fixed azido-AHA-transformed MCF-7 cells. In addition to protein labeling, glycans were also metabolically modified with azido-bearing artificial monosaccharide (Gal-NAz) to probe the universality of this method. Likewise, GalNAz-treated cells (MCF-7 and NIH-3T3) were selectively tagged with alkyne-terminated CPNs. Subsequently, they modified the carboxyl-functionalized ultrabright PFBT CPNs with STA and IgG to perform specific cell imaging [15].

CPE-antibody conjugates were prepared by means of direct bioconjugation between an antibody (CD 3 or CD 20) and PPE derivative (PPE 8 or PPE 9) having blue or red fluorescent emission [24]. PPE 8-CD 3 showed excellent specificity toward T-cells (Jurkat) and PPE 9-CD 20 selectively bound to B-cells (SUDHL-4). It was demonstrated that rationally designed conjugated polyelectrolytes can be covalently attached directly to an antibody as a fluorescent reporter molecule without affecting the recognition specificity of the antibody. Moreover, the developed CPEs and the convenient direct bioconjugation strategy are readily applicable to other biological molecules.

Besides generally known molecular targets, CPEs can also be used for specific cellular organelle imaging. Nilsson and co-workers employed conformation-sensitive PT derivatives to accomplish targeted imaging of lysosome-related acidic vesicles in cultured primary cells [25]. Because conformation-sensitive PT derivatives possessed the property of binding different entities with different emission colors, a potential capability of imaging various subcellular structures was indicated. More recently, a water soluble polythiophene with tyrosine kinase inhibitor lapatinib as side chain moieties was designed and synthesized. Together with its fluorescent characteristic and low cytotoxicity, PTL can be used for cell membrane imaging of living cells by targeting the intracellular domain of transmembrane proteins (Fig. 3.3) [26].

CPEs have also been used for fluorescent imaging of microorganisms. In 2008, Wang and co-workers reported the assembly of cationic PMNT (2) into microtubes by using negatively charged fungi (A. niger) as the template, polymer 2 could image the growth state of living fungi [27]. In 2009, Giorgetti et al. in situ fabricated fluorescent PPE derivative coated-silver NPs (Ag-CPNs) for sensing fungus P. variotii [28]. In terms of the interactions with P. variotii, the Ag-CPNs

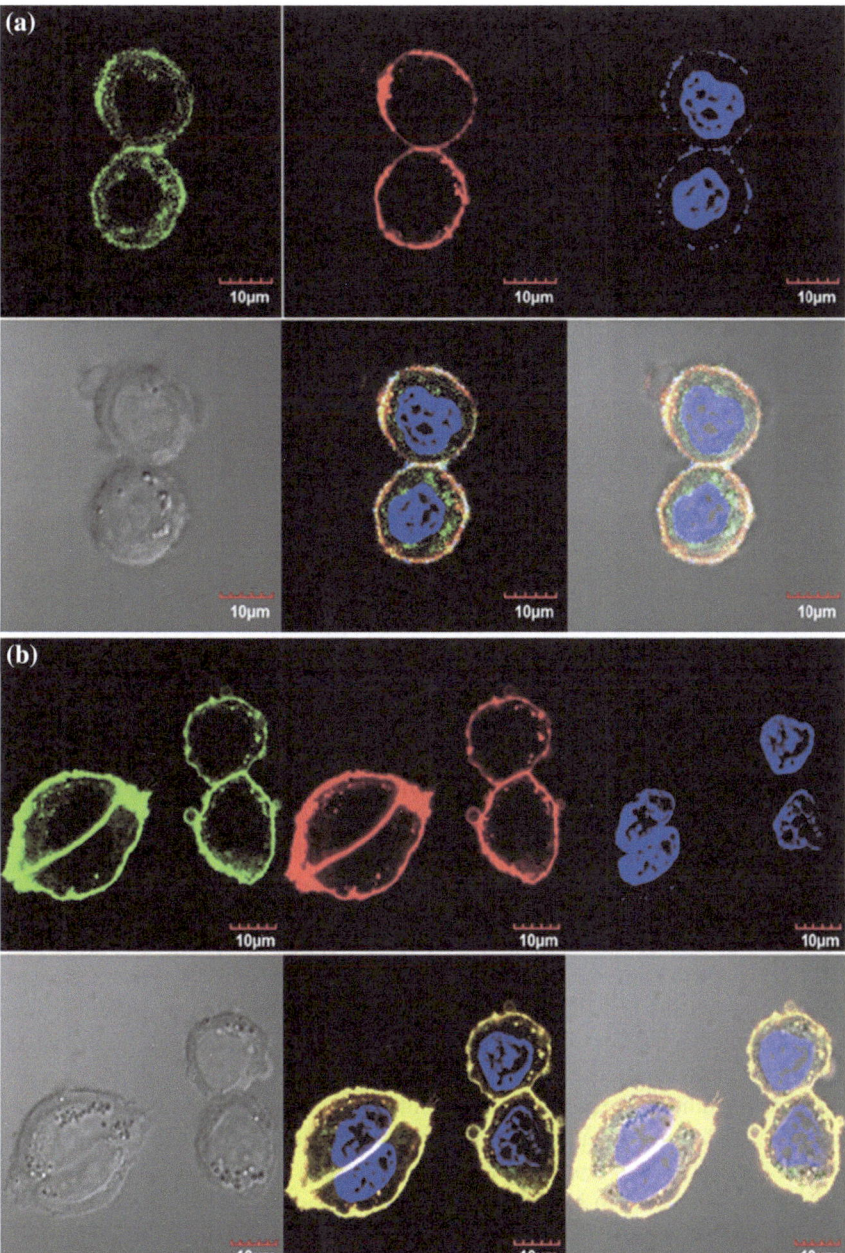

**Fig. 3.3** Cellular location of PTL in different cell lines after 48 h of incubation: **a** SK-BR-3 cell line, **b** MCF-7 cell line. The false colors of PTL, Dil (cell membrane stain), and Hoechest 33258 (nucleus stain) are *green*, *red*, and *blue* respectively. The merged color of *green* and *red* is *yellow* (Reprinted with permission from Ref. [26]. Copyright (2013) Royal Society of Chemistry)

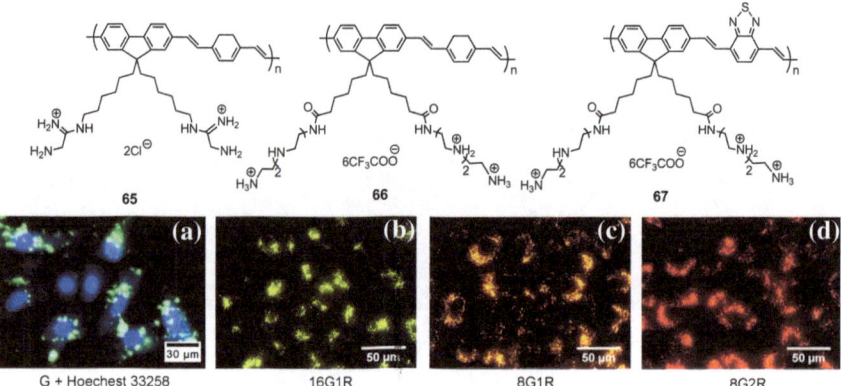

**Fig. 3.4** **a** Overlay image of A549 cells incubated with 66 and Hoechest 33258 dye, and fluorescent images of A549 cells incubated with barcoded microparticles 16G1R (**b**), 8G1R (**c**), and 8G2R (**d**) (where numbers denote the relative molar amount and G and R correspond to 66 and 67, respectively) (Reprinted with permission from Ref. [29]. Copyright (2012) Wiely-VCH Verlag GmBH & Co. KGaA Weinheim)

showed strong binding to both spores and mycelium to image the whole fungus, whereas the pure polymer only exhibited an imaging ability toward mycelium.

To overcome the small Stokes shifts of most organic dyes and achieve the simultaneous imaging with different emission colors under the same excitation source, Wang and co-workers employed microorganisms as the building blocks to fabricate multicolor microparticles by facile electrostatic and hydrophobic assembly of different CPEs and further realized cell imaging and optical barcoding [29]. By regulating intermolecular multistep FRET on noninvasive *E. coli* among three CPEs (blue-emissive 65, green-emissive 66, and red-emissive 67), 12 encoding colors were obtained. More importantly, the 65/66/67/*E. coli* microparticles with a Stokes shift of 170 nm could be excited under different excitation sources (408, 488, 532, and 645 nm) that are commonly equipped in most fluorescence instruments (e.g., CLSM and FCM). Further investigation indicated that the low-toxic barcoded micro-particles could be used for multicolor cell imaging (Fig. 3.4) and multiplexed flow analysis.

As a fundamental exploration, imaging with CPEs in vitro applications have been successful investigated. Further research focuses on imaging applications of CPEs ex/in vivo biological processes. Nilosson and co-workers have dedicated numerous contributions for real-time in vivo imaging of protein aggregates with PT derivatives [30]. The deposition of protein aggregates in various parts of human body gives rise to several devastating diseases, and the development of probes for the selective detection of aggregated proteins is crucial to advance our understanding of the pathogenesis underlying these diseases. PTs are fluorescent probes that bind selectively to protein aggregates. The conjugated thiophene backbone is flexible and offers a connection between the conformation and the emission properties, hence binding of PTs gives the molecule a spectral fingerprint.

Building on the selective labeling cells in vitro with surface-functionalized ultrabright CPNs, Chiu and co-workers further employed the recognition system and achieved in vivo tumor targeting [31]. The CPNs were composed of functional PSMA and deep-red emissive conjugate polymer blends (PFBT as the donor and PF-DBT5 as the acceptor) (Fig. 3.5). After covalent modification with a medulloblastoma-specific peptide chlorotoxin (CTX) and PEG, the photostable and serum-stable CTX-CPNs were subjected to malignant brain tumor imaging in a transgenic mouse model (ND2:SmoA1) via tail vein administration. As shown in Fig. 3.5, the fluorescence images of mouse brains taken after 72 h of injection, the preponderant accumulation was observed only for the group of ND2:SmoA1/CTX-CPNs, suggesting the specific binding. Histological analysis showed the accuracy of targeting imaging. The biodistribution profile of CTX-CPNs at 72 h postinjection exhibited remarkable uptake in the liver, and a tiny

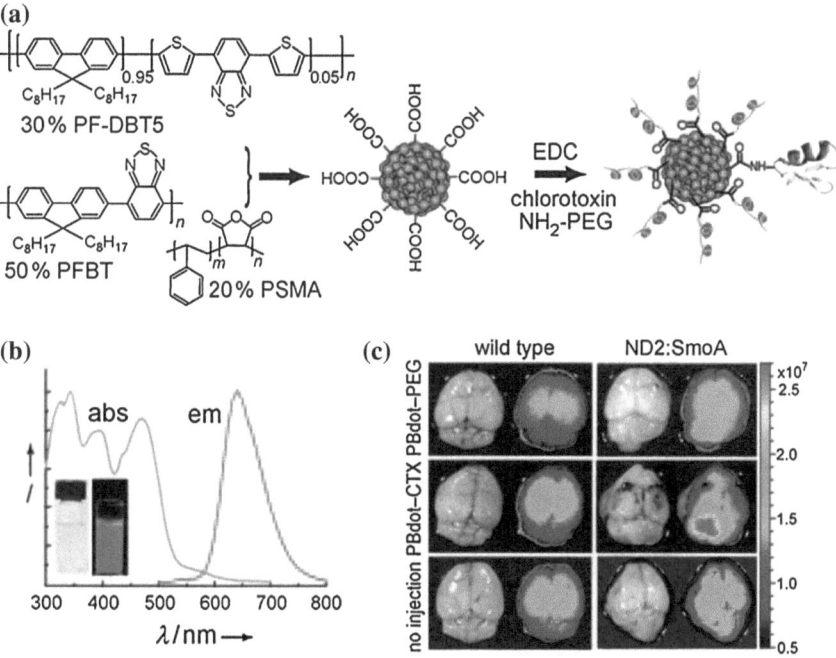

**Fig. 3.5** **a** CPN functionalization and CTX conjugation. A lightharvesting polymer PFBT, a red-emitting polymer PF-DBT5, and a functional polymer PSMA were co-condensed to form highly fluorescent CPNs with surface carboxyl groups. **b** Absorption and emission spectra of CPNs. Inset: photographs of an aqueous PBdot solution under illumination with ambient light (*left*) and UV light (*right*). **c** Fluorescence imaging of healthy brains in wild-type mice (*left*) and medulloblastoma tumors in ND2:SmoA1 mice (*right*) acquired at 72 h postinjection. Each mouse was injected with either nontargeting PEG-CPN (*top*) or targeting CTX-CPN (*middle*); the control is no injection (*bottom*). The *color gradient bar* corresponds to the fluorescence intensity (p/s/cm$^2$/sr) of the images (Reprinted with permission from Ref. [31]. Copyright (2011) Wiely-VCH Verlag GmBH & Co. KGaA Weinheim)

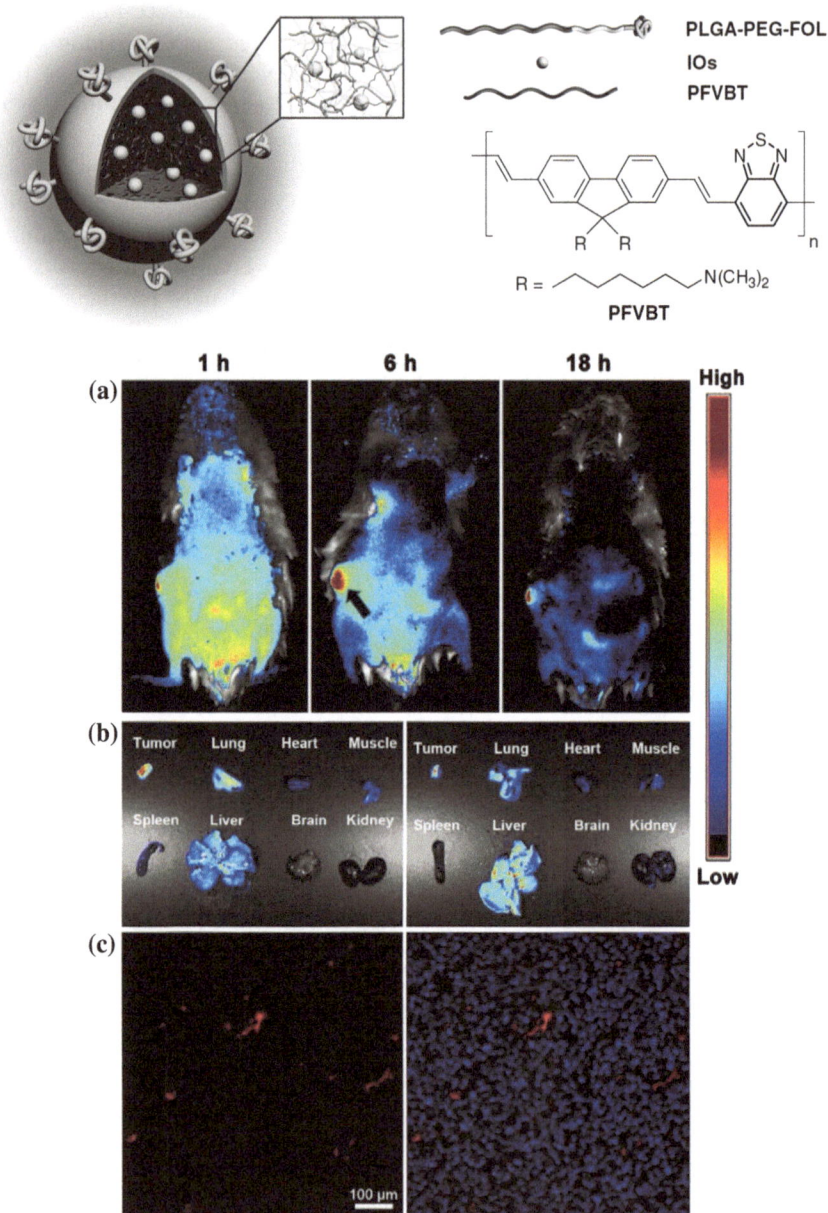

**Fig. 3.6** Illustration of the conjugated polymer based fluorescent-magnetic NP and the chemical structure of PFVBT (*up*). **a** Representative in vivo fluorescence images of mouse injected with FMCPNPs acquired at 1, 6 and 18 h post-injection. **b** Fluorescence images of various organs from the mice treated with FMCPNPs (*left*) and MCPNPs (*right*), respectively at 12 h post-injection. **c** Confocal images of sectioned tumor tissue harvested at 6 h post-injection. *Left image* shows the fluorescence of FMCPNPs in tumor section and *right image* shows the same image merged with 4',6-diamidino-2-phenylindole stained nuclei (*blue*) (Reprinted with permission from Ref. [33]. Copyright (2012) Wiely-VCH Verlag GmBH & Co. KGaA Weinheim)

portion of distribution in the spleen, almost no fluorescence signals in kidney, which indicated that the excretion route was from liver into bile and feces.

For in vivo imaging and drug tracking, Liu and co-workers prepared cisplatin-loaded NPs by linking PEG-grafted red-fluorescent CPNs with cisplatin via a ligand exchange reaction [32]. In PBS buffer, the CPE-PEG-Pt displayed a slow but persistent behavior of cisplatin releasing. Upon incubating the particles with HepG2 cancer cells in vitro, bright red fluorescence spreading the entire fixed cells could be observed, which demonstrated that CPE-PEG-Pt could be internalized by cells and enter into the nuclei. The CPE-PEG-Pt was injected into BALB/c nude mice through tail veins. Noninvasive live animal fluorescence imaging indicated the capability of the established NIR fluorescent probe to image in vivo. During the whole experimental process, the region of the liver was the main distribution site of PT-CPNs. The time-dependent signal decrease inferred that the CPE-PEG-Pt could be cleared away from the body to a certain extent. Moreover, the information of drug distribution could be simultaneously obtained by monitoring the fluorescence signals. In this case, the liver accumulation of CPE-PEG-Pt made the probe to be a potential agent to treat liver cancer. Very recently, Liu and co-workers fabricated dual-modal fluorescent-magnetic NPs with folate receptor over-expressed cancer cell targeting ability by co-encapsulation of PFVBT and lipid-coated iron oxides (IOs) into a mixture of PLGA-PEG-FOL and PLGA (Fig. 3.6) [33]. The emission of the PFVBT was designed to fall into FR/NIR spectral window. Incorporation of lipid-coated IOs into these NPs yielded probes having super paramagnetic properties without sacrificing their fluorescence. In vitro studies revealed that these dual-modal NPs can serve as an effective fluorescent probe to achieve targeted imaging of MCF-7 breast cancer cells without obvious cytotoxicity. In vivo fluorescence and magnetic resonance imaging results suggest that the NPs are able to preferentially accumulate in tumor tissues to allow dual-modal detection of tumors in a living body (Fig. 3.6). This work provides new opportunities to design conjugated polymer-based multimodal platform through incorporating various reagents of specific functionalities (e.g., therapeutic drugs and siRNA), which holds great promises for various in vivo bioimaging and therapy applications.

Switchable or activatable optical probes are unique in the field of molecular imaging since these agents can be turned on in specific environments but otherwise remain undetectable. This improves the achievable target-to-background ratios, enabling the detection of small tumors against a dark background. Therefore, switchable CPEs-based systems are promising to be designed and utilized to cell imaging with enhanced sensitivity.

# References

1. Celli JP, Spring BQ, Rizvi I, Evans CL, Samkoe KS, Verma S, Pogue BW, Hasan T (2010) Imaging and photodynamic therapy: mechanisms, monitoring, and optimization. Chem Rev 110:2795–2838
2. McRae RL, Phillips RL, Kim I-B, Bunz UHF, Fahrni CJ (2008) Molecular recognition based on low-affinity polyvalent interactions: selective binding of a carboxylated polymer to fibronectin fibrils of live fibroblast cells. J Am Chem Soc 130:7851–7853

3. Zhu C, Yang Q, Liu L, Wang S (2011) A potent fluorescent probe for the detection of cell apoptosis. Chem Commun 47:5524–5526
4. Pecher J, Mecking S (2010) Nanoparticles of conjugated polymers. Chem Rev 110:6260–6279
5. Moon JH, McDaniel W, MacLean P, Hancock LE (2007) Live-cell-permeable poly (p-phenylene ethynylene). Angew Chem Int Ed 46:8223–8225
6. Rahim NAA, McDaniel W, Bardon K, Srinivasan S, Vickerman V, So PTC, Moon JH (2009) Conjugated polymer nanoparticles for two-photon imaging of endothelial cells in a tissue model. Adv Mater 21:3492–3496
7. Howes P, Green M, Levitt J, Suhling K, Hughes M (2010) Phospholipid encapsulated semiconducting polymer nanoparticles: their use in cell imaging and protein attachment. J Am Chem Soc 132:3989–3996
8. Pu K-Y, Li K, Liu B (2010) Cationic oligofluorene-substituted polyhedral oligomeric silsesquioxane as light-harvesting unimolecular nanoparticle for fluorescence amplification in cellular imaging. Adv Mater 22:643–646
9. Tang F, He F, Cheng H, Li L (2010) Self-assembly of conjugated polymer-Ag@SiO$_2$ hybrid fluorescent nanoparticles for application to cellular imaging. Langmuir 26:11774–11778
10. Parthasarathy A, Ahn H-Y, Belfield KD, Schanze KS (2010) Two-photon excited fluorescence of a conjugated polyelectrolyte and its application in cell imaging. ACS Appl Mater Interfaces 2:2744–2748
11. Wu C, Bull B, Szymanski C, Christensen K, McNeill J (2008) Multicolor conjugated polymer dots for biological fluorescence imaging. ACS Nano 2:2415–2423
12. Wu C, Bull B, Christensen K, McNeill J (2009) Ratiometric single-nanoparticle oxygen sensors for biological imaging. Angew Chem Int Ed 48:2741–2745
13. Kim I-B, Shin H, Garcia AJ, Bunz UHF (2007) Use of a folate-PPE conjugate to image cancer cells in vitro. Bioconjugate Chem 18:815–820
14. Pu K-Y, Shi J, Cai L, Li K, Liu B (2011) Affibody-attached hyperbranched conjugated polyelectrolyte for targeted fluorescence imaging of her2-positive cancer cell. Biomacromolecules 12:2966–2974
15. Wu C, Schneider T, Zeigler M, Yu J, Schiro PG, Burnham DR, McNeill JD, Chiu DT (2010) Bioconjugation of ultrabright semiconducting polymer dots for specific cellular targeting. J Am Chem Soc 132:15410–15417
16. Ding D, Pu K-Y, Li K, Liu B (2011) Conjugated oligoelectrolyte-polyhedral oligomeric silsesquioxane loaded pH-responsive nanoparticles for targeted fluorescence imaging of cancer cell nucleus. Chem Commun 47:9837–9839
17. Li K, Pan J, Feng S-S, Wu AW, Pu K-Y, Liu Y, Liu B (2009) Generic strategy of preparing fluorescent conjugated-polymer-loaded poly(DL-lactide-co-Glycolide) nanoparticles for targeted cell imaging. Adv Funct Mater 19:3535–3542
18. Li K, Liu Y, Pu K-Y, Feng S-S, Zhan R, Liu B (2011) Polyhedral oligomeric silsesquioxanes-containing conjugated polymer loaded plga nanoparticles with trastuzumab (herceptin) functionalization for her2-positive cancer cell detection. Adv Funct Mater 21:287–294
19. Pu K-Y, Li K, Liu B (2010) Multicolor conjugate polyelectrolyte/peptide complexes as self-assembled nanoparticles for receptor-targeted cellular imaging. Chem Mater 22:6736–6741
20. Li K, Zhan R, Feng S-S, Liu B (2011) Conjugated polymer loaded nanospheres with surface functionalization for simultaneous discrimination of different live cancer cells under single wavelength excitation. Anal Chem 83:2125–2132
21. Zambianchi M, Maria FD, Cazzato A, Gigli G, Piacenza M, Sala FD, Barbarella G (2009) Microwave-assisted synthesis of thiophene fluorophores, labeling and multilabeling of monoclonal antibodies, and long lasting staining of fixed cells. J Am Chem Soc 131:10892–10900
22. Kandel PK, Fernando LP, Ackroyd PC, Christensen KA (2011) Incorporating functionalized polyethylene glycol lipids into reprecipitated conjugated polymer nanoparticles for bioconjugation and targeted labeling of cells. Nanoscale 3:1037–1045
23. Wu C, Jin Y, Schneider T, Burnham DR, Smith PB, Chiu DT (2010) Ultrabright and bioorthogonal labeling of cellular targets using semiconducting polymer dots and click chemistry. Angew Chem Int Ed 49:9436–9440

24. Lee K, Lee J, Jeong EJ, Kronk A, Elenitoba-Johnson KSJ, Lim MS, Kim J (2012) Conjugated polyelectrolyte-antibody hybrid materials for highly fluorescent live cell-imaging. Adv Mater 24:2479–2484
25. Bjork P, Nilsson KPR, Lenner L, Kagedal B, Persson B, Inganas O, Jonasson J (2007) Conjugated polythiophene probes target lysosome-related acidic vacuoles in cultured primary cells. Mol Cell Probes 21:329–337
26. Wang B, Zhu C, Liu L, Lv F, Yang Q, Wang S (2013) Synthesis of a new conjugated polymer for cell membrane imaging by using an intracellular targeting strategy. Polym Chem. doi: 10.1039/C3PY00097D
27. Liu L, Duan X, Liu H, Wang S, Li Y (2008) Microorganism-based assemblies of luminescent conjugated polyelectrolytes. Chem Commun 45:5999–6001
28. Giorgetti E, Giusti A, Arias E, Moggio I, Ledezma A, Romero J, Saba M, Quochi F, Marceddu M, Gocalinska A, Mura A, Bongiovanni G (2009) In situ production of polymer-capped silver nanoparticles for optical biosensing. Macromol Symp 283–284:167–173
29. Feng X, Yang G, Liu L, Lv F, Yang Q, Wang S, Zhu D (2012) A convenient preparation of multi-spectral microparticles by bacteria-mediated assemblies of conjugated polymer nanoparticles for cell imaging and barcoding. Adv Mater 24:637–641
30. Klingstedt T, Nilsson KPR (2012) Luminescent conjugated poly- and oligo-thiophenes: optical ligands for spectral assignment of a plethora of protein aggregates. Biochem Soc Trans 40:704–710
31. Wu C, Hansen SJ, Hou Q, Yu J, Zeigler M, Jin Y, Burnham DR, McNeill JD, Olson JM, Chiu DT (2011) Design of highly emissive polymer dot bioconjugates for in vivo tumor targeting. Angew Chem Int Ed 50:3430–3434
32. Ding D, Li K, Zhu Z, Pu K-Y, Hu Y, Jiang X, Liu B (2011) Conjugated polyelectrolyte-cisplatin complex nanoparticles for simultaneous in vivo imaging and drug tracking. Nanoscale 3:1997–2002
33. Li K, Ding D, Huo D, Pu K-Y, Thao NNP, Hu Y, Li Z, Liu B (2012) Conjugated polymer based nanoparticles as dual-modal probes for targeted in vivo fluorescence and magnetic resonance imaging. Adv Funct Mater 22:3107–3115

# Chapter 4
# Biomacromolecule Delivery System Based on Functionalized Conjugated Polyelectrolytes

**Abstract** In this chapter, biomacromolecule delivery systems on the basis of cationic conjugated polyelectrolytes are introduced. Cationic conjugated polyelectrolytes can form electrostatic complex with nucleic acids via electrostatic interactions, therefore cationic CPEs can be employed as gene carriers. By virtue of the self-luminous property, cationic CPEs can track the location of gene delivery and transfection in a real-time manner. Furthermore, cationic dendritic conjugated polyfluorene has been designed and used as DNA delivery vector with high transfection efficiency due to their excellent photostability and high quantum yield in water. CPEs have been developed as drug delivery systems due to their sturdy hydrophobic backbones, which are also presented.

**Keywords** Cationic CPEs • Gene delivery • Transfection reagent • Dendritic conjugated polyelectrolytes • Drug delivery and release

## 4.1 Gene Delivery

CPEs have established themselves as useful imaging agents for live cell due to their high fluorescence brightness, good photostability, and low toxicity. By virtue of the self-luminous property, cationic CPEs can be employed as gene carriers for real-time tracking the location of gene delivery and transfection. Cationic CPEs can form electrostatic complex with nucleic acids which facilitates genes binding, encapsulation, efficient cellular uptake, and effectively protecting genes against nuclease degradation. Moreover, much reduced cytotoxicity compared to commercially available transfection reagent (e.g., polyethylenimine (PEI) and lipofectamine), and no potential immune response compared to viral vectors are advantages offered by CPEs.

In 2008, Choi and co-workers realized in vitro gene delivery by using cationic PDA nanovesicles [1]. The cationic PDA nanovesicle was prepared by UV-initiated photopolymerization of amphiphilic cationic diacetylene monomer DAMDPA-bis-PCDA, and

S. Wang and F. Lv, *Functionalized Conjugated Polyelectrolytes*, SpringerBriefs in Molecular Science, DOI: 10.1007/978-3-642-40540-2_4, © The Author(s) 2013

the average diameter of PDA nanovesicle was 267.4 ± 25.1 nm. By means of electrostatic interactions, the cationic PDA nanovesicle enabled to form electrically neutral complexes with negatively charged plasmid DNA at the charge ratio of 4.6:1 (PDA to DNA). Cell viability analysis of PDA nanovesicles toward human embryonic kidney (HEK) 293 cells revealed that reduced cytotoxicity was obtained for PDA nanovesicles than commercially available PEI. In vitro transfection experiments in HEK 293 cells indicated that PDA nanovesicles at the charge ratio of 4.6:1 displayed an acceptable level of gene expression. Moreover, the transfection efficiency was further improved with increasing the charge ratio of PDA to DNA. Both charge ratios and transfection efficiency of the polymerized vesicles were higher than that of the nonpolymerized vesicles, which presumably arose from the fact that the stiff polymerized vesicles were easier to escape from endosomes so as to release the loaded plasmid DNA. Nevertheless, the transfection efficiency for PDA nanovesicles was much lower than that of PEI.

A lipid-modified cationic poly(fluorenylenephenylene) (PFPL, 68) was prepared and the amphiphilic PFPL itself formed uniform ca. 50 nm NPs in water with a QY of 6.3 % [2]. The design strategy was based on the fact that the lipid side chain may improve the biocompatibility of CPEs, and provide protective layers to CPE backbones, which make CPEs enter cytoplasm more easily. The obtained PFPL-NPs exhibited good photostability and low cytotoxicity, which was attributed to the protective effect of the outer layered hydrophobic and biocompatible lipid. In addition, PFPL-NPs were efficiently internalized by A549 cells to image the cells with

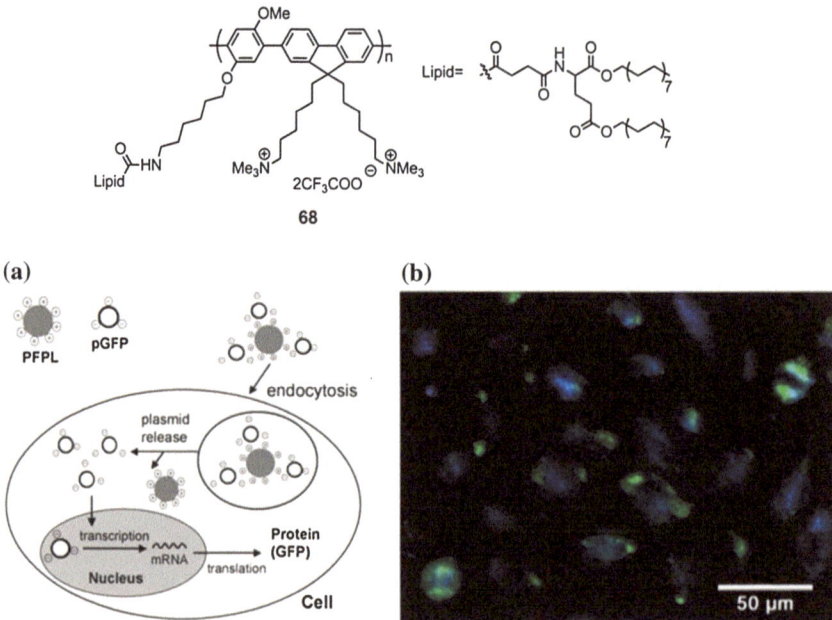

**Fig. 4.1 a** Schematic representation of PFPL-CPNs for cell transfection. **b** The overlap fluorescence image of A549 cells incubated with PFPL-CPN/pCX-EGFP complex (Reprinted with permission from Ref. [2]. Copyright (2010) Royal Society of Chemistry)

a distribution in the cytoplasm especially around the nuclear area. By employing PFPL-NPs as gene vector, DNA plasmid encoding green fluorescent protein (GFP) was effectively delivered into A549 cells, which was shown from the bright blue fluorescence. Moreover, the emergence of the green GFP fluorescence verified that the transcription and translation processes occurred (Fig. 4.1).

The amphiphilic characteristics of CPEs often lead to different aggregation structures in polar solvents, which results in fluorescence quenching due to π-stacking between the backbones of CPEs. Dendritic conjugated polyelectrolytes may overcome this drawback since the regular and repeated branched structures can prevent the aggregation of the polymers due to the steric hindrance. Cationic dendritic conjugated polyfluorene (CCP, 69) was designed and synthesized and examined as DNA plasmid delivery vector [3]. The obtained cationic dendritic conjugated polyfluorene exhibited excellent photostability and highly fluorescent with a quantum yield of 43 % in water. Figure 4.2 shows the phase contrast and fluorescence images of Hela cells treated with CCP and DNA plasmid complex. Both the blue fluorescence of conjugated polyfluorene and the green fluorescence of GFP were observed which suggested that CCP successfully delivered

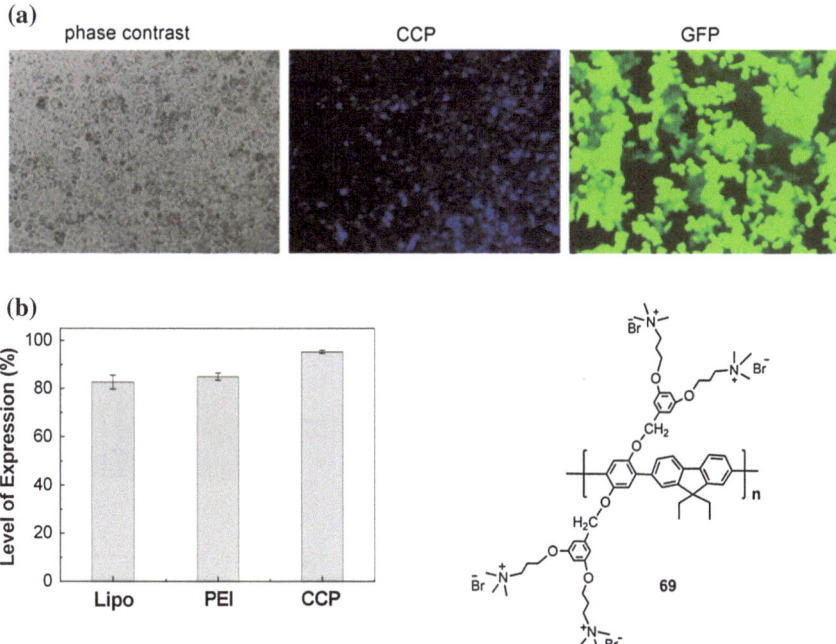

**Fig. 4.2  a** Phase contrast (*left*) and fluorescence microscopy images of Hela cells incubated with CCP/pDNA complex. Fluorescence image of CCP (*middle*) was recorded on fluorescence microscopy using a 380/30 nm excitation filter and that of GFP (*right*) was recorded using a 455/70 nm excitation filter. **b** Quantifying the expression of GFP using flow cytometry. Each bar represents the mean GFP expression in three independent experiments. (Reprinted with permission from Ref. [3]. Copyright (2012) Wiley–VCH Verlag GmBH & Co. KGaA Weinheim)

plasmid into cells for further transcription and translation. GFP expression experiments revealed that dendritic conjugated polyfluorene achieved efficient delivery and transfection of pDNA encoding GFP gene with 92 % efficiency (Fig. 4.2), which surpassed that of commercial transfection agents (lipofectamine and PEI). Furthermore, the cationic dendritic conjugated polyfluorene was utilized for multidrug resistance gene-targeted siRNA delivery in doxorubicin (Dox)-resistant human cervix carconima (MCF-7) cells. As a siRNA transfection agent, cationic dendritic conjugated polyfluorenes efficiently achieved the reversal of drug resistance and enhance drug sensitivity. The chemical structure imparted CPEs with new features and capabilities, which represented a major step toward designing and applying CPEs that function in both imaging and therapeutic applications.

## 4.2  Drug Delivery and Release

The sturdy hydrophobic backbones of CPEs can be utilized to associate and support cargos, therefore, CPEs have been developed as drug delivery system which are able to monitor drug loading, delivery, and release processes in a noninvasive and real-time manner (by fluorometer or simply by naked eyes under UV light).

Wang and co-workers developed a fluorescent DNA/SP-PPV hybrid hydrogel material for monitoring drug release [4]. DNA/SP-PPV hybrid hydrogel was prepared via in situ Wessling polymerization where salmon DNA was used as template. Cationic SP-PPV formed crosslinking complexes with negatively charged salmon DNA due to electrostatic attraction, and the complexes exhibited highly stability to DNase I digestion, heat and ultrasound. Simultaneously, the expected fluorescent property of SP-PPV was well kept in the resultant hydrogels.

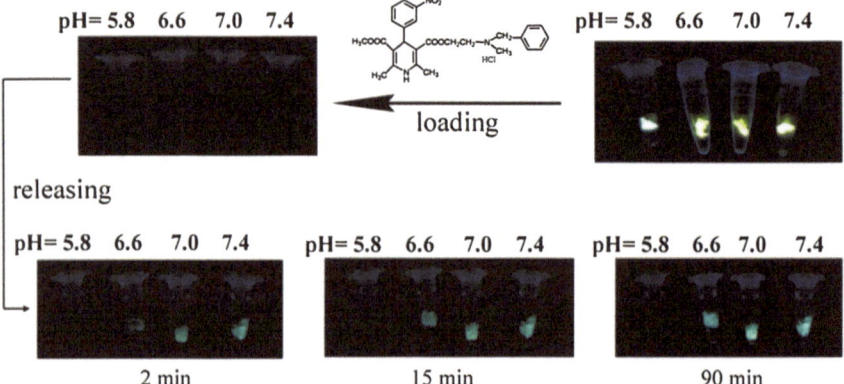

**Fig. 4.3** Monitoring the releasing of drug from Gel by observing changes of emission intensity of Gel after immersing into phosphate buffer solutions with different pH (5.8, 6.6, 7.0, and 7.4) at 2, 15, and 90 min (Reprinted with permission from Ref. [4]. Copyright (2009) Royal Society of Chemistry)

Lyophilizing wet gel could remove the occupied water molecules and result in forming porous material, which allows for loading drugs by diffusion. The chosen model drug (nicardipine hydrochloride) was an efficient electron-transfer quencher, so the loading process was accompanied with the fluorescence quenching of the SP-PPV gel. Consequently, the process of drug release could be tracked in real time by monitoring the recovery of fluorescence signals. The drug release investigation under various pH values (5.8–7.4) suggested that the release rate was pH-dependent (Fig. 4.3).

On the basis of SP-PPV based drug-release system, a CPE-based drug delivery and simultaneous drug-release monitoring system was developed by Wang and co-workers [5]. Cationic PFO formed electrostatic complexes with negatively

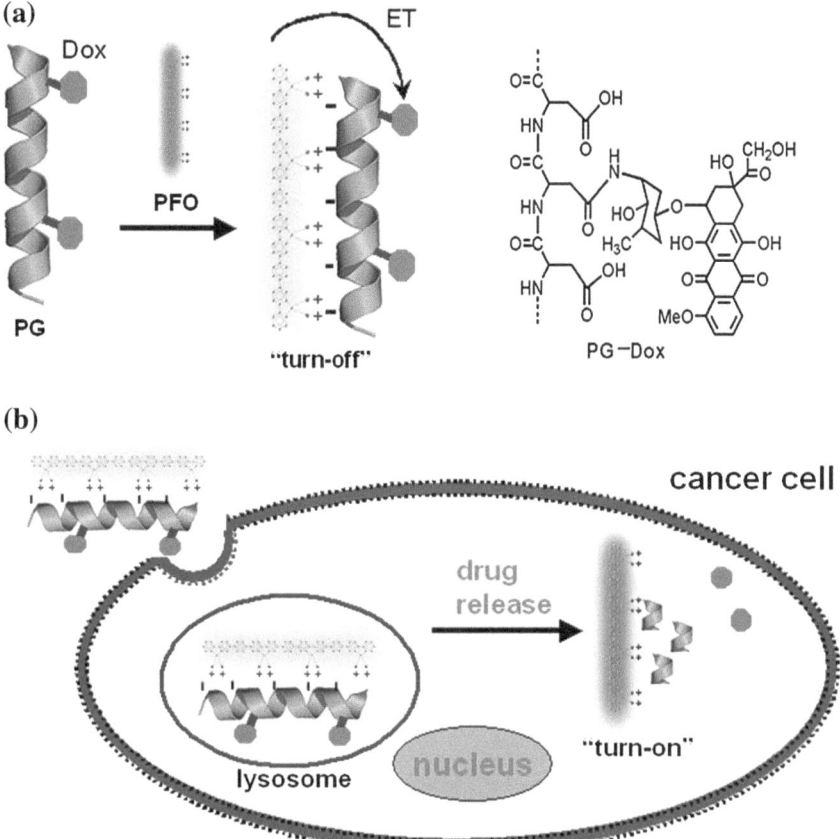

**Fig. 4.4 a** Schematic illustration of PFO/PG-Dox complexes, where the fluorescence of PFO is highly quenched by Dox. **b** Schematic representation of uptake of the electrostatic complexes into cancer cells. The hydrolysis of poly(L-glutamic acid) by hydrolase in lysosomes releases Dox to induce the activation of PFO fluorescence to "turn-on" state, thereby sensing the intracellular delivery of Dox

charged poly(L-glutamic acid) that was conjugated with anticancer drug Dox (PG-Dox). PFO exhibited good fluorescence quantum yield of 24 % in water, photostability, and little cytotoxicity. The fluorescence of PFO in PFO/PG-Dox complex was quenched by Dox via an electron-transfer mechanism, therefore the PFO/PG-Dox complex was in the fluorescence "turn-off" state. In the presence of carboxypeptidase or various hydrolase in lysosomes, the fluorescence recovery of PFO was observed due to the Dox release that was driven by the hydrolysis of poly(Lglutamic acid). The system converted to fluorescence "turn-on" state (Fig. 4.4). Furthermore, monitoring of drug release in A549 cancer cells was also achieved by using PFO/PG-Dox complexes.

A cationic pentathiophene (5T) was designed and synthesized which selectively accumulated in mitochondria to exhibit specific organellar imaging [6]. Meanwhile, 5T efficiently induced cell apoptosis associating with JNK pathway activation. 5T formed nanoparticles with sodium chlorambucil (a widely used anticancer drug) through electrostatic interactions, and the nanoparticles could deliver

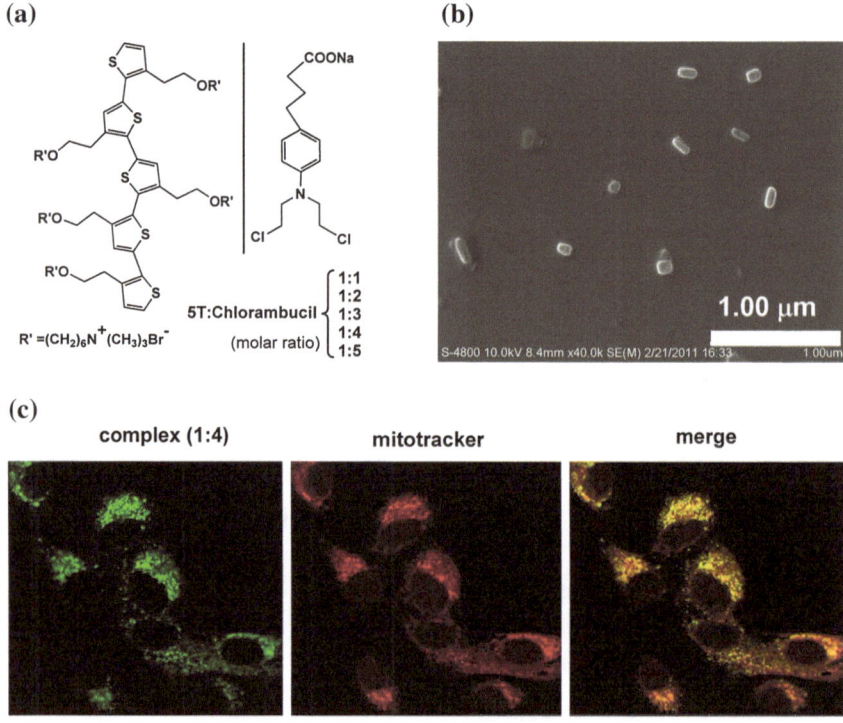

Fig. 4.5  **a** The 5T-chlorambucil complexes with varying molar ratio from 1:1 to 1:5. **b** Scanning electron microscopy image of 5T-chlorambucil complex (1:4). **c** Fluorescence images of A498 cells stained with mitochondrial dye, complex (1:4), and the overlapped image. The representative image of complex (1:4) is *green*, the organelle-specific dye is *red* and the overlapped image is *yellow* (Reprinted with permission from Ref. [6]. Copyright (2012) Wiley–VCH Verlag GmBH & Co. KGaA Weinheim)

sodium chlorambucil to the mitochondria (Fig. 4.5). In addition, 5T/chlorambu-cil nanoparticles presented 2–9 fold more cytotoxicity than free 5T and sodium chlorambucil itself, which suggested that synergistic effects were obtained for the combination of 5T and chlorambucil.

# References

1. Yu GS, Choi H, Bae YM, Kim J, Kim J-M, Choi JS (2008) Preparation of cationic polydia-cetylene nanovesicles for in vitro gene delivery. J Nanosci Nanotechnol 8:5266–5270
2. Feng X, Tang Y, Duan X, Liu L, Wang S (2010) Lipid-modified conjugated polymer nanoparti-cles for cell imaging and transfection. J Mater Chem 20:1312–1316
3. Feng X, Lv F, Liu L, Yang Q, Wang S, Bazan GC (2012) A highly emissive conjugated poly-electrolyte vector for gene delivery and transfection. Adv Mater 24:5428–5432
4. Tang H, Duan X, Feng X, Liu L, Wang S, Li Y, Zhu D (2009) Fluorescent DNA-poly(phenylenevinylene) hybrid hydrogels for monitoring drug release. Chem Commun 45:641–643
5. Feng X, Lv F, Liu L, Tang H, Xing C, Yang Q, Wang S (2010) Conjugated polymer nanoparti-cles for drug delivery and imaging. ACS Appl Mater Interfaces 2:2429–2435
6. Yang G, Liu L, Yang Q, Lv F, Wang S (2012) A multifunctional cationic pentathiophene: Synthesis, organelle-selective imaging, and anticancer activity. Adv Funct Mater 22:736–743

# Chapter 5
# Drug Screening Applications of Functionalized Conjugated Polyelectrolytes

**Abstract** In this chapter, three examples are illustrated for drug screening applications of conjugated polyelectrolytes. First, on the basis of the AChE detection system, the screening of inhibitors has been developed with high sensitivity by virtue of signal amplification property of CPEs. Second, specific RNA–protein interaction can be prevented by potential drugs, and FRET efficiency between CPE and fluorescein (labeled on protein) can be reduced to different extents by small organic molecules. Third, a FRET-based method to evaluate antimicrobial susceptibility has been developed. And integrated with an automated microplate reader, the established assay method could be used for high-throughput antibiotic screening.

**Keywords** Drug screening • Acetylcholinesterase activity • Enzyme inhibitor • RNA–protein interaction • Förster resonance energy transfer

Signal amplification property of CPEs imparts remarkably improved sensitivity for chemical and biological detection schemes based on optical methods. Thus, CPEs are very suitable to be employed for drug screening such as evaluation of the inhibition effect of enzyme inhibitors, ascertaining antibiotics susceptibility, and high-throughput screening of potential drugs.

As many enzymes are targets for active drugs, drug screening is an important application of CPE-based enzyme assays. A sensitive fluorescence turn-on assay for acetylcholinesterase (AChE) activity was developed where reversible fluorescence quenching of PBS-PFP by dabcyl-labeled acetylcholine [1]. Electrostatic attraction between PBS-PFP and Ach-dabcyl led to quenching of PBS-PFP. After adding AChE to the sensing system, Ach-dabcyl was hydrolyzed to generate choline and a negatively charged dabcyl moiety. The dabcyl moiety was repulsed by PBS-PFP and moved away from the electrostatic complex, and then the quenched fluorescence of PBS-PFP was recovered, as illustrated in Fig. 5.1. On the basis of the AChE detection system, the screening of inhibitors can be simultaneously

S. Wang and F. Lv, *Functionalized Conjugated Polyelectrolytes*, SpringerBriefs in Molecular Science, DOI: 10.1007/978-3-642-40540-2_5, © The Author(s) 2013

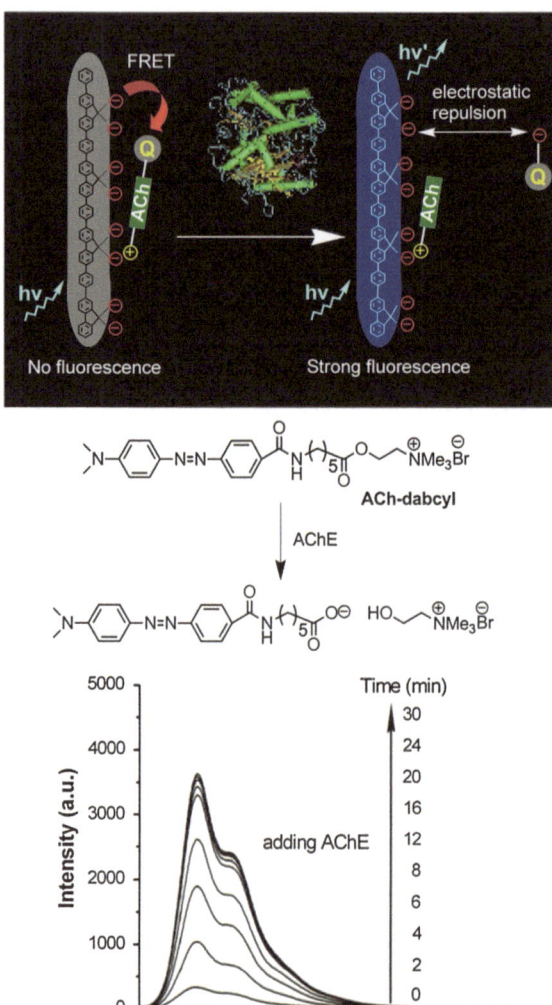

**Fig. 5.1** Schematic illustration of AChE activity assays and the corresponding hydrolysis of Ach-dabcyl by AChE (*up*). Emission spectra of PBS-PFP/Ach-dabcyl as a function of reaction time for AChE-catalyzed hydrolysis (*down*) (Reprinted with permission from Ref. [1]. Copyright (2007) Wiley–VCH Verlag GmBH & Co. KGaA Weinheim)

achieved. Galanthamine and donepezil, clinical inhibitors for the treatment of AD, were used as model drugs, and the inhibitor concentration required to reach 50 % enzyme activity ($IC_{50}$) was measured. The obtained $IC_{50}$ values were 12 and 23 nM for galanthamine and donepezil, respectively.

Specific RNA–protein interactions may mediate the replication cycles of pathogenic viruses (e.g., HIV-1 and picornaviruses), therefore small organic molecules that target viral RNA sites and prevent RNA–protein complexation can be considered as potential candidates for drug discovery [2]. The regulatory peptide (Rev) binding to the Rev responsive element (RRE) sequence in the RNA of HIV-1 was chosen as a model RNA/peptide binding pair. The design and anticipated function of the PFP/Rev-Fl/RRE system was diagrammatically shown in Fig. 5.2. The fluorescein-labeled arginine-rich Rev peptide (positively charged at neutral pH values) complex with negatively charged RRE forms a Rev-Fl/RRE complex with net negative charge. The addition of PFP resulted in the formation of PFP/Rev-Fl/RRE electrostatic complexes, in which PFP resides in close proximity to fluorescein, and efficient FRET took place. Neomycin B is a small-molecule antibiotic commonly used as a competitive inhibitor toward the RRE/Rev complex, which can bind to the major groove of duplex RNA through electrostatic interactions. When neomycin was titrated to the PFP/RRE/Rev-Fl solution, the three-way binding equilibrium shifts toward the release of free Rev-Fl. This process resulted in the removal of Rev-Fl from the vicinity of PFP and a concomitant reduction in FRET efficiency. In comparison with kanamycin, as shown in Fig. 5.2, neomycin B led to a larger reduction in FRET efficiency, which was attributed to the relatively efficient enzyme inhibition ability of neomycin B.

Wang and co-workers developed a FRET-based rapid, simple, and sensitive method to evaluate antimicrobial susceptibility and further demonstrate the ability of the system to screen antibiotics in a high-throughput manner (Fig. 5.3) [3].

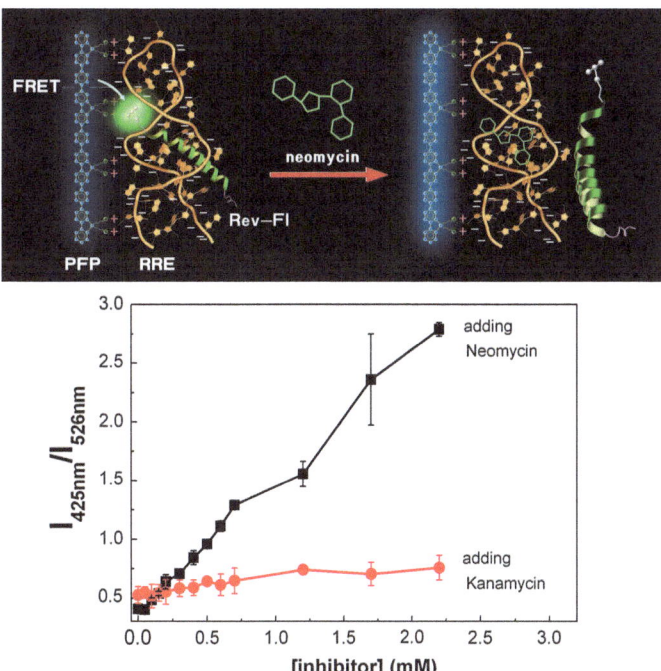

**Fig 5.2** Schematic representation of the mechanism for the detection of neomycin inhibition (*up*). FRET ratio as a function of the neomycin B and kanamycin concentration (*down*) (Reprinted with permission from Ref. [2]. Copyright (2009) Wiley–VCH Verlag GmBH & Co. KGaA Weinheim)

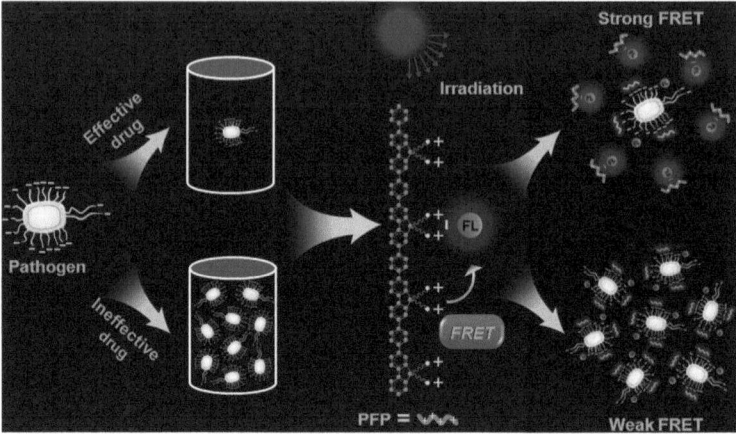

**Fig. 5.3** Schematic representation of antimicrobial susceptibility assessment

In the presence of ineffective or effective antibiotics, negatively charged bacteria exhibited exponential proliferation or growth suppression, respectively. Efficient FRET between PFP 5 and fluorescein can be regulated to different extents by the addition of bacteria treated with effective or ineffective antibiotics through competitive electrostatic interactions. Facilitated by the signal-amplifying of CPEs, the detection limit of the *E. coli* amount was as low as $10^4$ CFU (colony forming unit), which was beneficial for greatly shortened detection time to 4–5 h). By monitoring the change of FRET signal, the minimum inhibitory concentration and $IC_{50}$ values of the tested antibiotics were obtained. More importantly, integrated with an automated microplate reader, the established assay method could be used for high-throughput antibiotic screening.

# References

1. Feng F, Tang Y, Wang S, Li Y, Zhu D (2007) Continuous fluorometric assays for acetylcholinesterase activity and inhibition with conjugated polyelectrolytes. Angew Chem Int Ed 46:7882–7886
2. An L, Liu L, Wang S, Bazan GC (2009) An optical approach for drug screening based on light-harvesting conjugated polyelectrolytes. Angew Chem Int Ed 48:4372–4375
3. Zhu C, Yang Q, Liu L, Wang S (2011) Rapid, simple, and high-throughput antimicrobial susceptibility testing and antibiotics screening. Angew Chem Int Ed 50:9607–9610

# Chapter 6
# Therapeutic Applications of Functionalized Conjugated Polyelectrolytes

**Abstract** Conjugated polyelectrolytes can be induced to generate reactive oxygen species upon light illumination, which act as photosensitizers in photodynamic therapy to damage pathogenic microorganisms and cancer cells. In this chapter, the light-activated biocidal activities of PPE derivatives are presented where PPE derivatives act as photosensitizers. A multifunctional cationic PPV 3 for simultaneously accomplishing selective recognition, imaging, and killing of bacteria over mammalian cells has been reported. Moreover, energy transfer systems have been illustrated to enhance light-activated antibacterial activity. CPEs are also exploited to function concurrently as an anticancer agent and an imaging reporter, and furthermore the anticancer specificity can be achieved through delicate molecular design. CPE-drug conjugates have been constructed for intracellular molecule-targeted binding and inactivation of protein for growth inhibition of cancer cells. Bioluminescence has been employed to replace light source in PDT through bioluminescence resonance energy transfer to conjugated polyelectrolytes, which opens a new therapy modality to tumor and pathogen infections.

**Keywords** Photodynamic therapy • Photosensitizer • Reactive oxygen species • Biocidal activity • Antitumor activity • Bioluminescence resonance energy transfer

## 6.1 Mechanism of Photodynamic Therapy

Photodynamic therapy (PDT) is a photochemistry-based approach that utilizes a light-activatable chemical, termed as a photosensitizer (PS), and light of an appropriate wavelength to impart cytotoxicity via the generation of reactive molecular species such as singlet oxygen ($^1O_2$). PDT is a well-established approach for cancer treatment, and the use of PDT to treat bacterial and fungal infections has been in practice for over 30 years [1, 2]. PDT can destruct microorganisms featured as the leakage of bacterial contents, and therefore bacteria will not readily

S. Wang and F. Lv, *Functionalized Conjugated Polyelectrolytes*, SpringerBriefs in Molecular Science, DOI: 10.1007/978-3-642-40540-2_6, © The Author(s) 2013

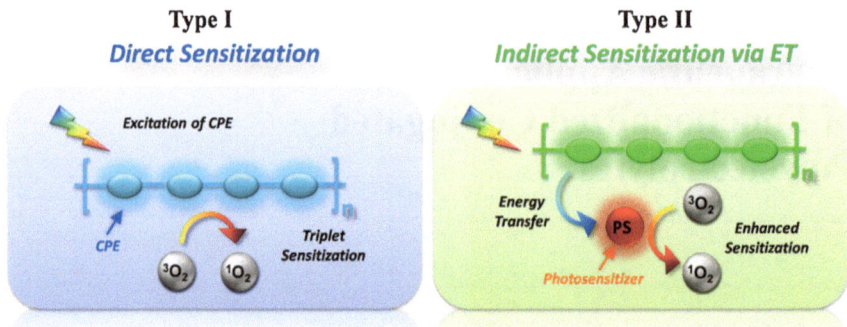

**Fig. 6.1**  Two types of oxygen sensitization manner provided by CPEs

develop resistance to PDT, which provide alternative antibacterial therapeutics for resolving the multi-antibiotic resistant issue. Most of the PS under investigation for the treatment of cancer and other tissue diseases are based on the tetrapyrrole nucleus. CPEs are found to be induced to generate reactive oxygen species upon light illumination, which can act as photosensitizers in PDT to damage pathogenic microorganisms and cancer cells (type I, Fig. 6.1). Besides direct sensitization of oxygen molecules by CPEs, another strategy (type II, Fig. 6.1) utilizing energy transfer from CPEs to available PS (such as porphyrin) to improve the $^1O_2$ quantum yield is appealing, which is facilitated by the large excitation cross-section of CPEs. In addition, quaternary ammonium (QA) groups were reported being capable to kill a variety of microorganisms, and the killing efficacy significantly enhanced with the increase of QA contents [3]. The biocidal mechanism was explained as the exchange of counterions between cationic QA groups and bacterial membranes, which results in cell death by the destabilization of cell membranes and the following release of cytoplasmic constituents. Therefore, CPEs functionalized with QA groups would exhibit effective biocidal capability against a broad spectrum of bacteria and spores in the dark. It should be noted that, if QA-CPEs possess light-killing capability, the total antimicrobial activity originated from the combination of light toxicity and dark toxicity.

## 6.2  Anti-Microorganism Activity

Under physiological conditions, microbial pathogens, including bacteria and fungi, are negatively charged. The Whitten group pioneered antimicrobial research with CPEs utilizing the electrostatic interaction between cationic CPEs and microorganisms. In 2005, they reported that a cationic PPE derivative 70 (chemical structure shown in Fig. 6.2) was used to coat and subsequently kill Gram-negative *E. coli* and Gram-positive *B. anthracis* in a visible-light induced manner [4]. They believed that the antimicrobial effects should be general and a range of conjugated

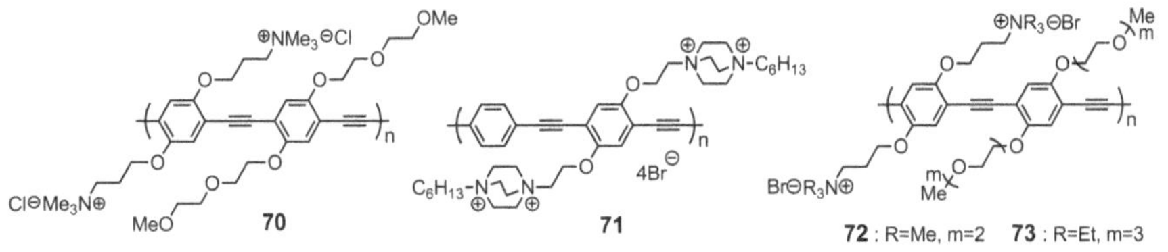

**Fig. 6.2** Chemical structures of CPEs 70–73

polyelectrolytes in different formulations would be developed as a useful new class of biocides for both dark and light-activated applications.

Along these lines, Whitten and co-workers further developed the light-activated biocidal activity of three cationic QA-containing PPE derivatives (71–73, chemical structures shown in Fig. 6.2) by supporting them on colloids [5]. Because PPE derivatives could be excited to a triplet state that sensitized the nearby oxygen molecules to generate singlet oxygen, the CPEs themselves displayed prominent light-induced biocidal effects toward Gram-negative *C. marina* and *P. aeruginosa*. To gain more insight into the biocidal mechanism, PPE derivatives (72, 73) were anchored on colloidal silica particles by simple physisorption or covalent graft. The resultant surface-grafted conjugated polyelectrolytes (SGCP) beads similarly displayed effectively light induced biocidal behavior against *C. marina* and *P. aeruginosa*. The proposed mechanism for the biocidal activity of the CPEs was illustrated in Fig. 6.3 with several sequential steps: (a) SGCP captured bacteria via electrostatic and hydrophobic interactions; (b–d) singlet oxygen and/or other subsequently produced ROS generated at the polymer/bacteria interface under irradiation, resulting in the damage of bacteria; (e) aggregates of dead bacteria and colloidal particles formed as a result of the agglomeration of bacterial debris.

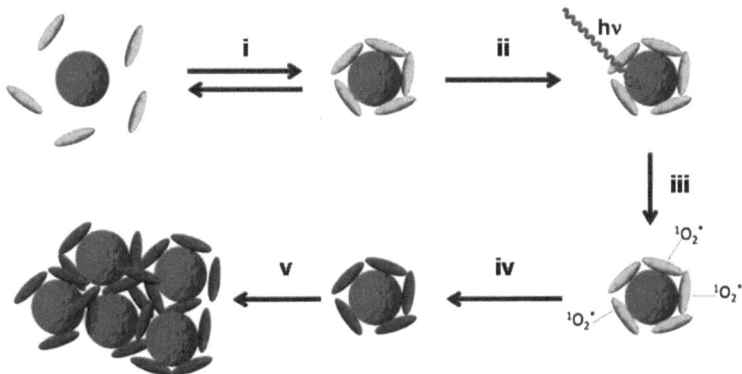

**Fig. 6.3** Mechanism of biocidal action of SGCP. **a** reversible bacteria adhesion to the particle; **b** photoexcitation of CPE; **c** singlet oxygen generation; **d** killing bacteria by singlet oxygen; **e** aggregation of particles

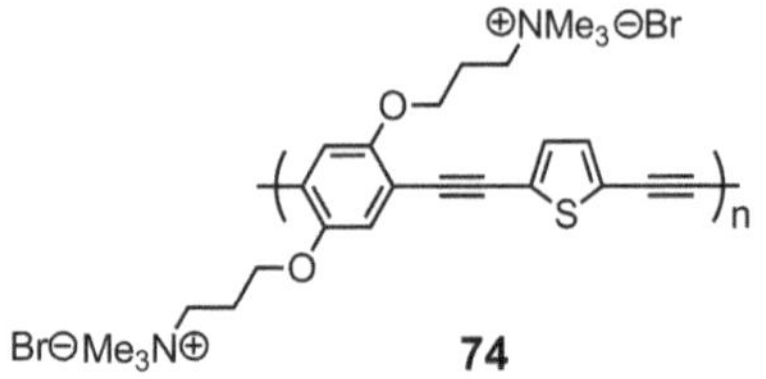

**Fig. 6.4** Chemical structure
of CPE 74

74

Whitten and co-workers investigated the differences regarding the light and dark biocidal activities of cationic PPEs (71–74) [6]. For PPE (71–73)-supported colloids mentioned above, light-activated biocidal activity against *P. aeruginosa* and *C. marina* were remarkable and little killing effect were observed in dark for a short time. Even over prolonged incubation, only a slow dark destruction was observed. Differently, PPE derivative 74 (chemical structure shown in Fig. 6.4) exhibited conspicuous dark biocidal activity against *P. aeruginosa* with an over 95 % killing efficiency and little enhanced antimicrobial performance under light irradiation. The efficient dark biocidal activity of PPE derivative 74 was originated from its highly lipophilic structure and its antimicrobial QA groups to destroy the outer membranes of microorganisms. The low light toxicity was ascribed to the higher aggregation both in solution and on the microspheres, which led to the poor capability in the generation of singlet oxygen and other ROS. These results revealed the mechanisms of dark toxicity and demonstrated that the aggregation of CPEs was harmful to light-activated biocidal activities, providing new insights into the regulation of killing mode.

Inspired by the findings that colloidal polymer-bead assemblies can entrap and kill bacteria, Whitten and co-workers further developed light-induced PPE antimicrobial microcapsules with high biocidal efficiency [7]. The PPE microcapsules were fabricated through alternately depositing four bilayers of anionic PPE 36 and cationic PPE 71 onto $MnCO_3$ template particles by layer-by-layer assembly followed by dissolution of template, leaving the microcapsules with a layer of positive charges in the exterior and a hollow hole in the interior. The special microstructures of the CPEs microcapsules imparted them with strong capability to irreversibly entrap and kill bacteria both within and at the surfaces with an over 95 % killing efficiency toward *P. aeruginosa* after exposure to 1 h of white light.

In 2010, Whitten and co-workers synthesized four "end-only" functionalized oligo(phenylene ethyneylene)s (EO-OPEs, 75–78, chemical structure shown in Fig. 6.5) and investigated their capability for killing bacteria [8]. Except anionic OPE 78 (due to Coulombic repulsion), the other three cationic OPEs exhibited efficient light-induced biocidal behavior toward both Gram negative *E. coli* and Gram-positive *S. epidermidis* and *S. aureus* under 365 nm irradiation. OPE 77 presented nearly 100 % killing efficiency due to the highest sensitized ability to generate singlet oxygen. In contrast to Gram-negative bacteria, the biocidal activity against Gram-positive bacteria was better, which gave rise to the more complicated cell wall compositions and protective mechanisms of Gram-negative microorganisms. Similar to CPEs-based biocidal process, cationic OPEs initially displayed strong

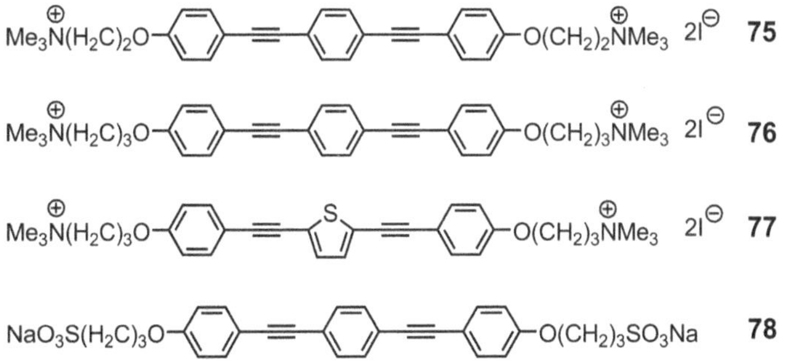

**Fig. 6.5** Chemical structures of CPEs 75–78

association with negatively charged bacterial membrane (presumably penetration) via electrostatic and hydrophobic interactions. Under irradiation, the generation of singlet oxygen and/or ROS at the interface of OPE-bacteria destroyed the normal cell metabolic activities, ultimately resulting in bacterial damage and death.

The antibacterial activity of symmetric and asymmetric cationic OPEs (79 and 80, chemical structure shown in Fig. 6.6) against Gram negative *E. coli* and Gram-positive *S. epidermidis* and *S. aureus* were explored both in the dark and under irradiation by UV light [9]. All OPEs exhibited efficient dark toxicity, and the biocidal effects increased with an increasing number of repeat units, concentrations, and incubation time. The symmetric oligomers 79 possessed stronger antibacterial activity than that of the asymmetric 80. For light-activated killing, 79-1 presented the highest biocidal effect in the 79 series as a result of the large singlet oxygen quantum yield, whereas 80-3 mediated the strongest killing efficiency in the 80 series due to the efficient membrane disruption ability and singlet

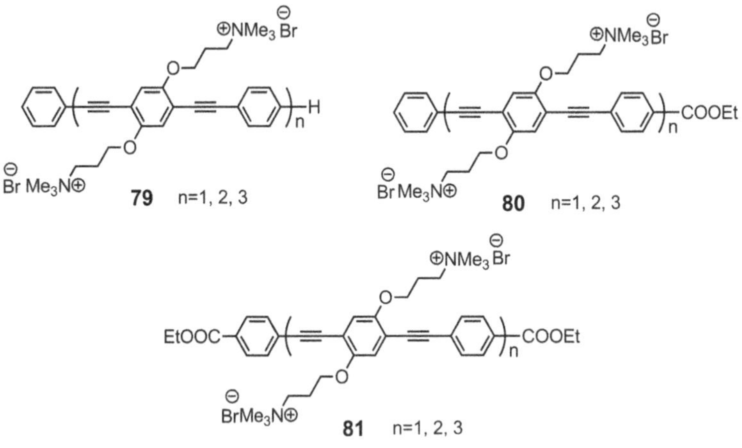

**Fig. 6.6** Chemical structures of CPEs 79–81

oxygen-sensitizing ability. Correspondingly, the proposed light-induced antibacterial mechanism was attributed to the combination of bacterial membrane disruption and the interfacial or intracellular generation of singlet oxygen or other reactive oxygen species, where the latter was the dominated factor for the biocidal activity.

Great deals of work have been carried out in Whitten group to get further insights into the biocidal mechanism of PPE and OPE-based cationic CPEs [10–12]. Different model anionic lipid membranes were utilized to simulate bacterial cell membranes, for dark-toxic polymer 74, it could rapidly and efficiently insert into lipid membranes, which accounted for the excellent dark antimicrobial activity. Lipid headgroup charge and membrane fluidity were considered as other factors to further probe the antimicrobial mechanism of PPE 74 [12]. PPE 74 showed selective binding and insertion into anionic phosphatidylglycerol lipid membranes as a result of electrostatic attraction, which was responsible for the main driven force to associate with negatively charged bacteria so as to favor the subsequent bacterial membrane disorganization. The membrane-perturbation capability of several cationic PPEs (36, 71, 72, and 74) and OPEs (76–78, 81) with different side-chains were investigated to better understand how these antimicrobial molecules interact with membranes [13]. There is no destructive effect toward lipid membranes was observed for anionic 36 and the shortest cationic 80-1 due to electrostatic repulsion or unbound attribute and the insufficient length to span the entire membrane, respectively. Because of the high charge density and hydrophobic alkyl chains of 71 and the comparable linear length of 76 and 77 to insert into the lipid bilayer, polymer 71 and two "end-only" functionalized OPEs (76 and 77) are able to perturb both model mammalian and bacterial membranes. Facilitated by the strong electrostatic driven forces, other cationic PPEs and OPEs exhibited pronounced membrane-disruption selectivity toward bacterial-membrane mimics over mammalian cell-membrane mimics. For OPEs, the longer the molecules were, the easier the bacterial membrane was destroyed. Building on aforementioned work, Whitten and co-workers employed different characterization techniques, from morphological to molecular scales to gain more complete understanding of both the light-activated and dark toxicity mechanisms for the CPE and OPE compounds. The structures of the antimicrobial agents and bacterial outer envelope control their interactions as well as the biocidal mechanisms, which provide an important guideline for further designing new class of biocidal agents.

Recently, Whitten and co-workers extended the utility of PPEs and OPEs to the antifungal and sporicidal activities in dark and under UV-irradiation against *S. cerevisiae* vegetative cells, germinated ascospores and asci [14]. In the dark, PPE-DABCO 71, EO-OPE-1 (DABCO) 76 and OPE-3 (80-3) exhibited comparable or higher antifungal activities compared to the widely used antibiotic AmB. With UV-irradiation, all the tested agents induced strong reductions in yeast cell viability, and the antifungal activities of the CPEs or OPEs were dependent on the growth phase of the yeast cells, where cells in growth phases that correspond to higher metabolic activities were more susceptible to the biocidal activities of CPEs and OPEs. Limited inactivation activities towards ascospores were obtained for all the tested CPEs and OPEs. In the dark, there were no effect for reducing spore viability, and only EO-OPE-1 77 was active under UV irradiation, which inactivated more than 95 % of the yeast ascospores. The CPEs and OPEs exhibited efficient activities against ascospores once they undergo germination under UV light irradiation. The protein-enriched outer envelope of yeast cells and germinated ascospores were revealed to be main target for the CPE and OPE antimicrobial materials.

As for the tested CPEs and OPEs above mentioned, the structure–reactivity relationships between their photophysical properties and antibacterial activity revealed that OPEs possess pronounced light activated biocidal activity. Moreover, photophysical studies indicated that OPEs were efficient sensitizers for singlet oxygen, which was believed to play an important role in their biocidal activity. However, the absorption of these studied OPEs was primarily in the near-UV or blue-violet region of the visible spectrum, which limited their utilization as light-activated biocides. Whitten and co-workers introduced donor–acceptor electronic structure in designing OPEs to make them absorb visible region of spectrum. The antibacterial activity against *Staphylococcus aureus* (*S. aureus*) of oligomers 82–84 (chemical structures shown in Fig. 6.7) were investigated with visible light as the illumination source. Terthiophene oligomer 82 exhibited the most efficient biocidal activity, which was consistent with the photophysical data showing that 82 had higher intersystem crossing and singlet oxygen sensitization efficiency. The introduction of donor–acceptor electronic structure represented a major step toward designing and applying OPES as light-activated biocides under the irradiation with visible light.

Wang and coworkers reported a multifunctional cationic PPV 3 for simultaneously accomplishing selective recognition, imaging, and killing of bacteria over mammalian cells [15]. In a mixture of bacteria and mammalian cells, PPV 3

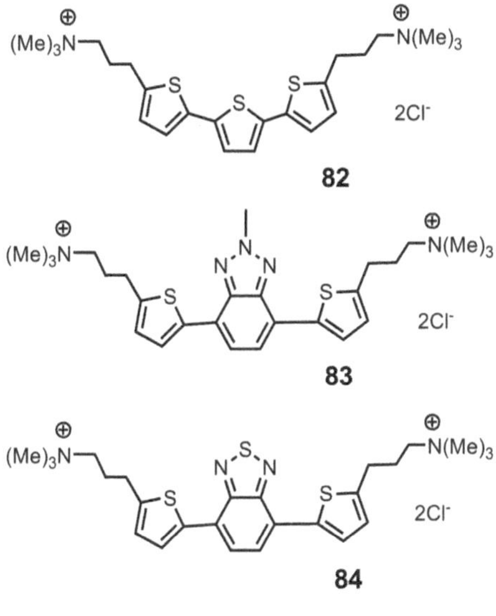

**Fig. 6.7** Chemical structures of CPEs 82–84

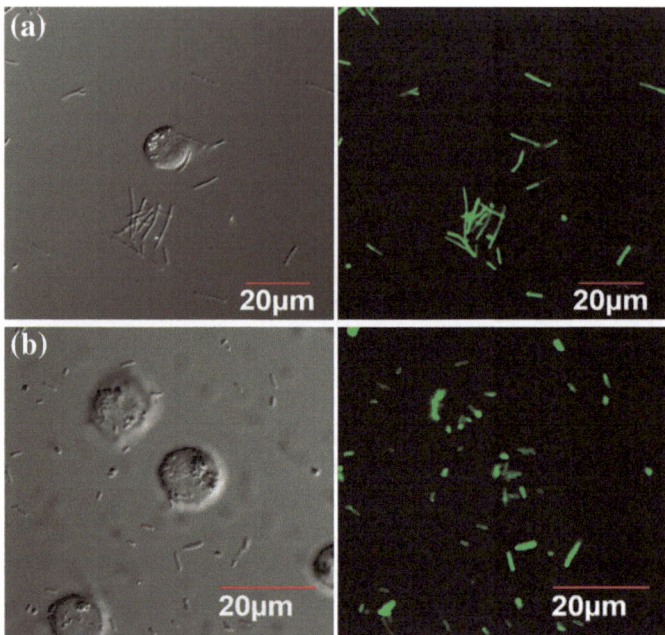

**Fig. 6.8** **a** CLSM images of Jurkat T cells and Gram-positive *B. subtilis* incubated with PPV 3 and **b** Jurkat T cells and Gram-negative Amp$^r$ *E. coli* incubated with PPV-3. *Left* bright field images, *right* fluorescent images (Reprinted with permission from Ref. [15]. Copyright (2011) Wiely-VCH Verlag GmBH & Co. KGaA Weinheim)

exhibited selective electrostatic binding toward bacteria due to the higher negative charges of bacterial surface over mammalian cells and further lit them up (Fig. 6.8). On the basis of the differential binding behaviors of PPV 3, the selective killing of bacteria was successfully achieved, where mammalian cell viability was almost intact. The antimicrobial activity of PPV 3 originated from the combination of dark-toxic QA groups and light-toxic PPV 3 main backbone. Although the biocidal efficiency toward Gram-negative *E. coli* under white light was somewhat modest (<70 %) compared to the aforementioned systems, this was the first example of CPEs that integrated recognition, imaging, and killing functions to achieve the collaborative and selective antimicrobial effect.

As indicated in Fig. 6.1, utilizing energy transfer from CPEs to available PS can improve the $^1O_2$ quantum yield due to the large excitation cross-section of CPEs. In 2009, Xu and co-workers. demonstrated that cationic PFP and anionic hematoporphyrin could form electrostatic complexes and an over ninefold enhancement of hematoporphyrin emission for one-photon excitation FRET (380 nm) and 30-fold enhancement for two-photon excitation FRET (800 nm) were observed [16]. By extending the conjugation length and coplanarity of CPEs, Xu and co-workers. synthesized CPEs incorporating ethynylene (polymer 85) and vinylene (polymer 86) (chemical structures shown in Fig. 6.9) bridges into the backbones of cationic PFP

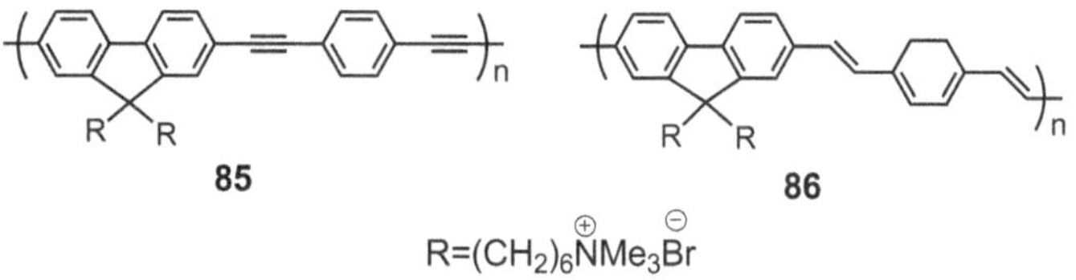

$$R=(CH_2)_6\overset{\oplus}{N}Me_3\overset{\ominus}{Br}$$

**Fig. 6.9** Chemical structures of CPEs 85 and 86

to improve the two-photon optical cross-sections of CPEs. Facilitated by the efficient FRET from CPEs to photosensitizer Rose Bengal, an 85-fold enhancement for two-photon excitation emission of Rose Bengal was obtained. Recently, Tang and co-workers developed CPEs/PS-doped mesoporous silica nanocomposite FRET system, where enhanced energy transfer was observed that offered by light-harvesting conjugated polymers [17]. Three porphyrin-based PSs were doped into the mesoporous silica nanoparticles to certificate the feasibility of this FRET model. All these studies results provide the foundation for the potential PDT application of CPEs-based energy-transfer systems.

PT-porphyrin energy transfer system was first illustrated to enhance light-activated antibacterial activity by Wang et al. [18]. By virtue of electrostatic interactions, anionic PT donor (87, chemical structure shown in Fig. 6.10) and cationic TPPN formed tight complex particles with net positive charges on the surfaces. Under white-light irradiation, efficient energy transfer through Dexter mechanism from 87 to TPPN followed by intersystem crossing to triplet, and sensitized oxygen molecules to produce singlet oxygen (Fig. 6.11). Owing to the excellent capability of light harvesting and light amplification of 87, the generation efficiency of singlet oxygen was enhanced than that direct excitation of the sensitizer TPPN. The positively charged 87/TPPN complexes exhibited efficient association with Gram-negative *E. coli* and Gram-positive *B. subtilis* driven by electrostatic and hydrophobic interactions. The photosensitized inactivation of bacteria for the PTP/TPPN complex was efficient, and about 70 % reduction of bacterial viability for *E. coli* and above 90 % reduction of bacterial viability for *B. subtilis* was observed after 5 min of irradiation with white light at a fluence rate of 90 mW (Fig. 6.11).

**Fig. 6.10** Chemical structure of CPE 87

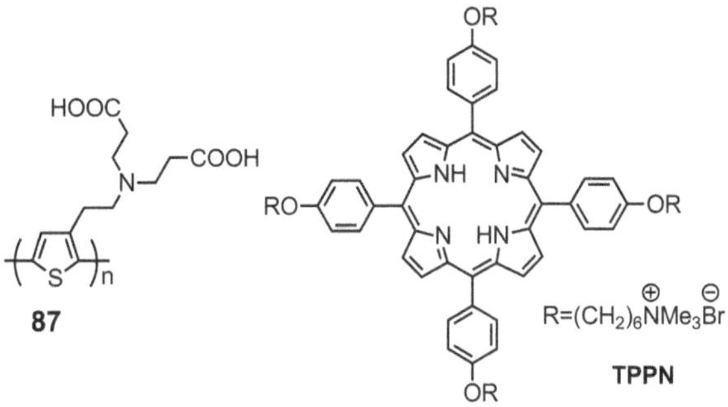

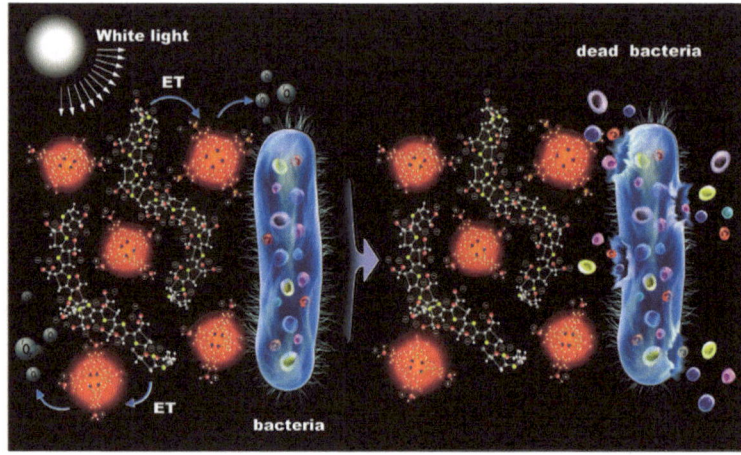

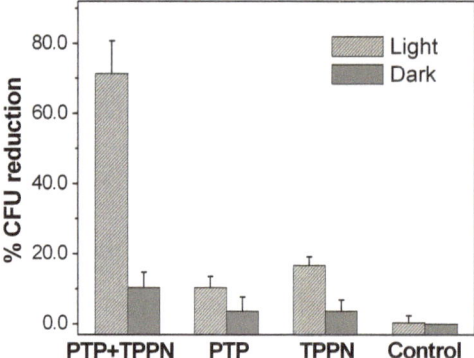

**Fig. 6.11** Schematic antibacterial mechanism of PTP/TPPN complex (*up*), and biocidal activity of PTP/TPPN, PTP, and TPPN toward *E. coli* in the dark and under white-light illumination for 5 min (*down*). Dark and light control experiments were done with the cell suspensions irradiated or in the dark in the absence of photosensitizers (Reprinted with permission from Ref. [18]. Copyright (2009) American Chemical Society)

By covalently attaching porphyrin moieties to the light-harvesting polythiophene backbone, the distance between the donor and acceptor was reduced for improving efficient energy transfer, thereby enhancing the efficiency of $^1O_2$ generation [19]. The resultant PT-porphyrin dyad (PTP 88) exhibited efficient light-activated antifungal activity in lower doses of irradiation light and polymer concentration. *A. niger* was chosen as model fungus in the antifungal experiments. Because the cell envelope of *A. niger* is mainly composed of chitin, glucans, polyphosphate, and a small quantity of proteins, lipids, and other biopolymers, electrostatic and hydrophobic interactions with the amphiphilic PTP 88 (chemical structure shown in Fig. 6.12) lead to strong binding [20]. Therefore, the generated $^1O_2$ are sufficiently close in proximity to effectively damage the outer membrane and inhibit spore germination.

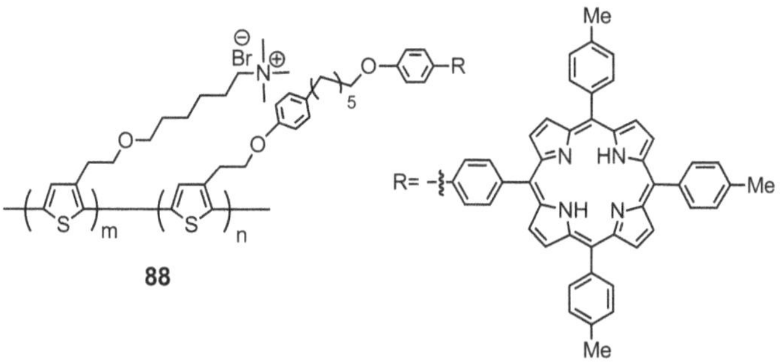

**Fig. 6.12** Chemical structure of CPE 88

The use of antimicrobial CPEs offers promise for enhancing the efficacy of some existing antimicrobial agents and minimizing the environmental problems accompanying conventional antimicrobial agents by reducing the residual toxicity of the agents, increasing their efficiency and selectivity, and prolonging the lifetime of the antimicrobial agents. Research concerning the development of antimicrobial conjugated polymers represents a great challenge for both the academic world and industry.

## 6.3 Anti-Tumor Activity

In addition to the antimicroorganism activity, the antitumor activity of CPEs has gained great interests and emerged as an important application of CPEs. In 2010, Wang et al. successfully achieved cancer cell killing and simultaneous apoptosis imaging utilizing cationic PMNT (polymer 2) [21]. PMNT was able to enter into the cells in a passive diffusion manner and localize in the cytoplasm, and induced cell apoptosis by the internalized PMNT, which was confirmed by the upregulated caspase-3. With the aid of irradiation light at 455/70 nm, cell apoptosis caused by PMNT was accelerated only within several minutes. Moreover, PMNT could be used for monitoring the apoptosis process by selectively staining apoptotic cells with a pattern of dense yellow clusters, which avoided adding the extra apoptosis-specific imaging agents. However, the specific target site of PMNT and the selective mechanism remained unclear which need to be further elaborated.

The aforementioned antifungal agent PTP 88 was exploited to function concurrently as an anticancer agent and an imaging reporter [22]. Two types of cancer cells, including A549 and A498 cells were chosen as the models to investigate the light-activated killing ability of PTP 88. As expected, upon irradiation at 470 nm for 30 min, an obvious decrease in cell viability for both cells was observed due to the cytotoxic singlet oxygen sensitized via the intermolecular energy transfer from PT backbones to porphyrin units. PTP 88 did not exhibit toxicity toward cells in the dark due to the low pendant porphyrin moieties. In order to attain the objective

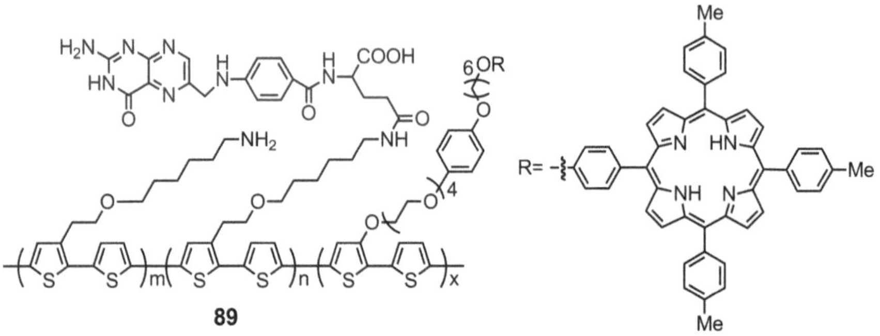

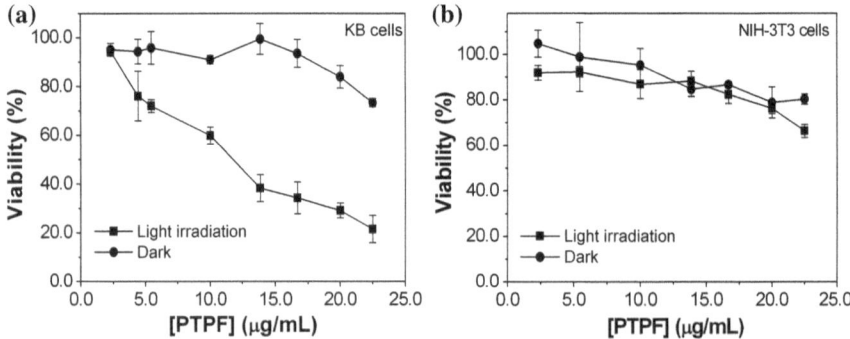

**Fig. 6.13** Chemical structure of folate-functionalized PT 89

of selectively recognition and killing of cancer cells, folate-functionalized PT 89 (chemical structure shown in Fig. 6.13) was synthesized to target folate receptor over-expressed cells. KB cancer cells with over-expressed folate receptor and folate receptor-negative NIH-3T3 fibroblasts cells were chosen as target cells to test the specificity of PT 89. A specific internalization was found in KB cells over NIH 3T3 cells after incubation with PT 89 for 24 h. Cell viability analysis indicated that the light-induced cytotoxicity of PT 89 was selective against KB cells with over 80 % cell damage, whereas for NIH 3T3 cells above 70 % cells survived (Fig. 6.14). This work demonstrates that anticancer specificity can be achieved through delicate molecular design, which indicated that novel polymers modified with molecular fragments will be designed and synthesized with specific anticancer capability.

Recently, Wang and co-workers designed and synthesized multifunctional CPE-drug conjugates for intracellular molecule-targeted binding and inactivation of protein for growth inhibition of cancer cells [23]. Estrogen-mediated growth of

**Fig. 6.14** Dose–response curves for cell viability of KB cells (**a**) and NIH-3T3 cells (**b**) treated with PT 89 by using a typical MTT assay under light irradiation or in the dark. *Error bars* represent standard deviations from three separate measurements (Reprinted with permission from Ref. [22]. Copyright (2011) Wiely-VCH Verlag GmBH & Co. KGaA Weinheim)

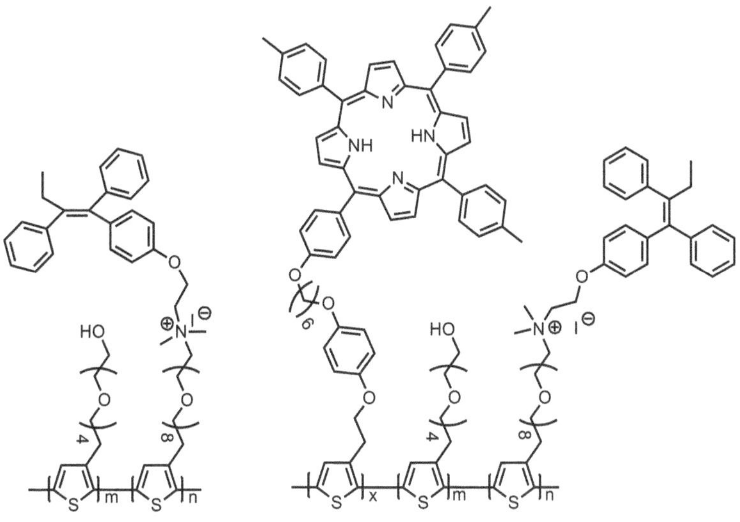

**Fig. 6.15** Chemical structures of PDT 90 and PTDP 91

human tumors (such as breast tumor MCF-7 cells) can be inhibited by inactivating estrogen receptor $\alpha$ (ER$\alpha$) using antiestrogen drugs. Tamoxifen (TAM), the most widely used estrogen receptor modulator [24–26], was linked to side chains of polythiophene to get polythiophene-tamoxifen conjugates PTD 90. Under white light irradiation, PTD 90 can sensitize oxygen to produce ROS that specifically inactivated the targeted ER$\alpha$ protein, retard the estrogen signal pathway and selectively inhibit the growth of the signal pathway relied breast tumor cells. Covalent attachment of porphyrin moieties to the light harvesting backbone of PTD yielded PTDP 91 (chemical structure shown in Fig. 6.15), which constrains interchromophore distances for optimizing energy transfer (ET) from polythiophene to porphyrin. Thus, PTDP significantly increased the ROS generation relative to PTD 90, and MCF-7 cells treated with PTDP 91 displayed quite remarkable sensitivity to light irradiation as expected (Fig. 6.16). Simultaneously, the fluorescent properties of polymers can serve to trace the cellular uptake and localization of polythiophene-drug conjugates by fluorescence imaging.

Cationic CPEs could form complex with *E. coli* by virtue of electrostatic and hydrophobic interactions as demonstrated above. Utilizing this property, Wang and co-workers developed a bacterial vector-based CPE system which exhibited efficient anticancer activity [27]. The preparation of bacterial vectors and their anticancer mechanism are demonstrated in Fig. 6.17. Red-emissive CPE was used for coating *E. coli.*, and facilitated by the collaborative release effect caused by CPE and a membrane disrupting antibiotic polymyxin B (PLB), CPE-coated *E. coli* were transported into the cells by endocytosis. Due to the excellent ROS sensitizing ability of various CPEs, CPE-coated *E. coli* was regarded as a ROS trigger

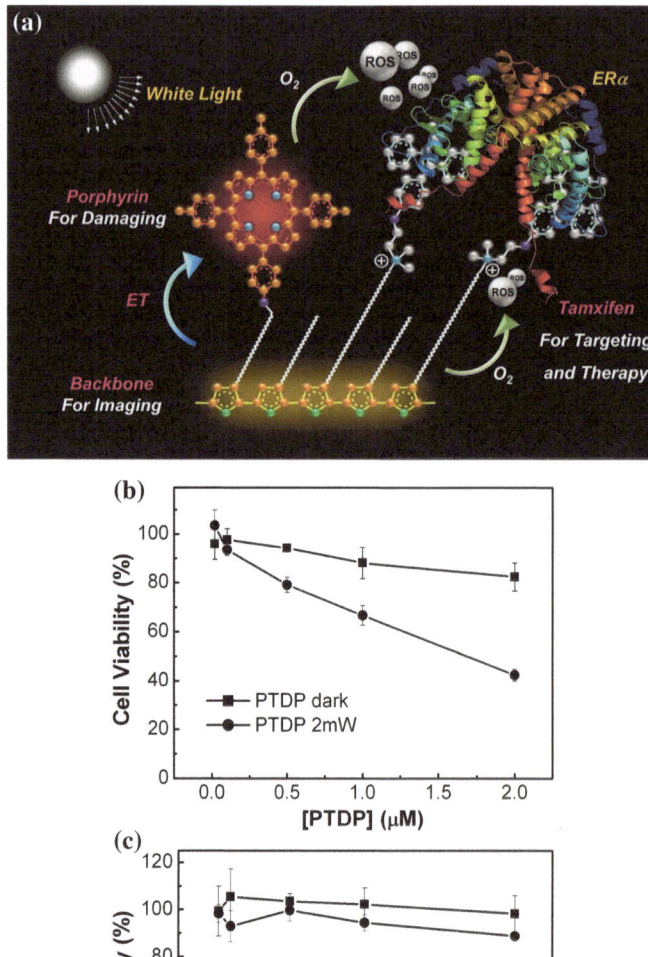

**Fig. 6.16** Schematic mechanism of PTDP for selective targeting and inactivation of intracellular estrogen signal pathway protein (**a**), Dose-response curves for cell viability of MCF-7 cells (**b**) and MDA-MB-231 cells (**c**) treated with PTDP 91 by using a typical MTT assay under light irradiation or in the dark (Reprinted with permission from Ref. [23]. Copyright (2012) Nature Publishing Group)

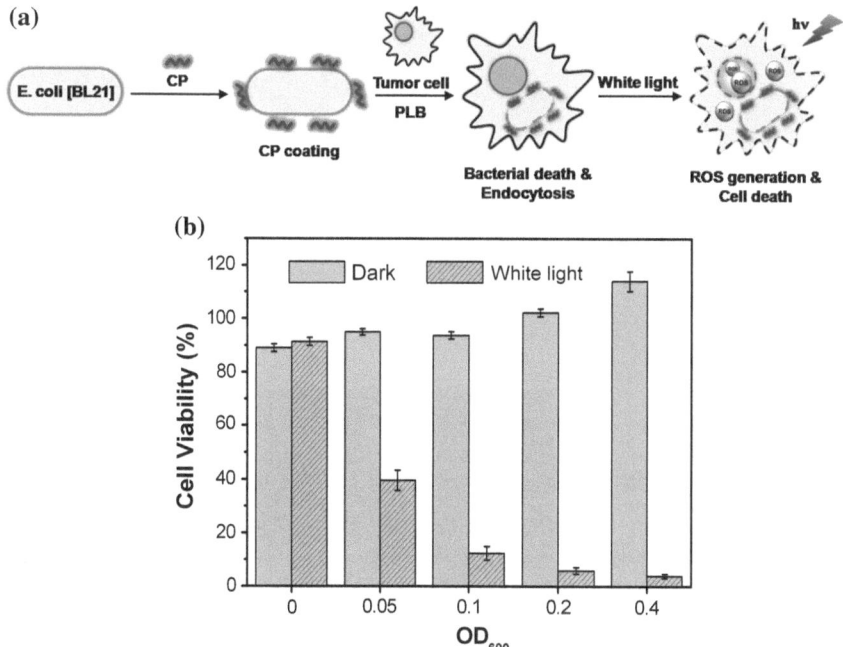

**Fig. 6.17** **a** Schematic diagram of cationic CPE coated live bacteria and intracellular killing of cancer cells. **b** Cell viability analysis of CPE coated *E. coli* against A498 cells under white light. Standard deviations are shown as *error bars* from five parallel experiments (Reprinted with permission from Ref. [27]. Copyright (2013) Wiely-VCH Verlag GmBH & Co. KGaA Weinheim)

source for light-mediated cell damage. Figure 6.17b presented that in the bacterial concentration range of 0–0.4 $OD_{600\ nm}$ (optical density at 600 nm), the killing efficiency increases dramatically and the highest killing efficiency can reach over 95 %. On the contrary, the dark groups are basically unaffected, which quantitatively proved that the CPE-coated bacteria were biocompatible.

The requirement of outer light source limits effective application of PDT to the lesions in deeper tissue. To eliminate this limitation, Wang et al. developed a novel PDT system for killing cancer in which the photosensitizer is activated by chemical molecules instead of light source (Fig. 6.19a) [28]. To demonstrate the concept, luminol, hydrogen peroxide, and horseradish peroxidase (HRP) were used as bioluminescent molecules and a cationic oligo (phenylene vinylene) (OPV, chemical structure shown in Fig. 6.18) was used as the photosensitizer. Bioluminescence resonance energy transfer (BRET) occurs between luminol and OPV since they meet the spectral overlap requirement. The excited OPV sensitizes the surroundings oxygen molecule to generate reactive oxygen species that kill the adjacent cancer cells in vitro and in vivo (Fig. 6.19). As shown in Fig 6.19b, less cytotoxicity was observed when the OPV concentration is less than 8 μM without luminol luminescence system. While in the presence of the luminol luminescence system,

**Fig. 6.18** Chemical structure of OPV

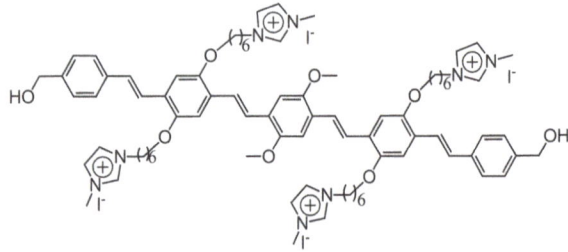

**Fig. 6.19** **a** Schematic illustration of the BRET system for PDT. **b** Cell viability of HeLa cells after incubation with OPV in the absence and presence of luminol luminescence system (E + S). **c** Tumor inhibition of mice post 18 days treatment, 10 mice used per group. Values are expressed as mean ± SD (n = 10, P < 0.001) (Reprinted with permission from Ref. [28]. Copyright (2012) American Chemical Society)

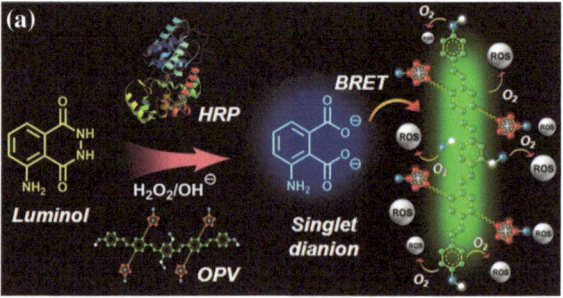

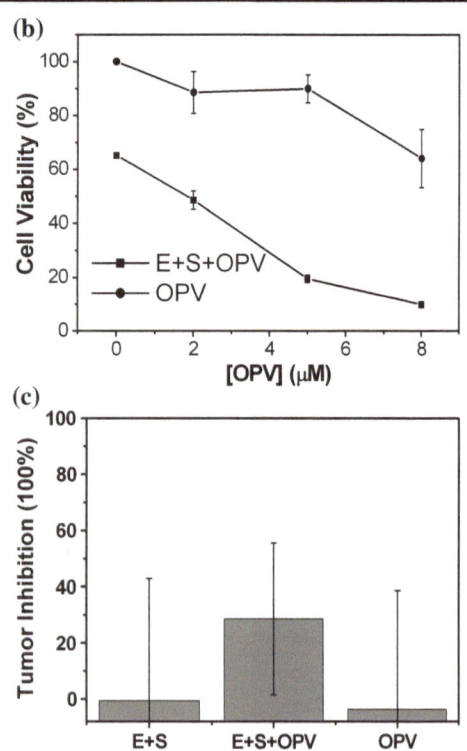

OPV displays more prominent cytotoxicity to Hela cells. The antitumor therapeutic efficacy of the novel BRET system was evaluated in HeLa cell tumor-bearing nude mice using intratumoral injection mode. A 55 % of tumor inhibition ratio was observed with the aid of luminal signal enhancer to prolong the light emission in BRET system (Fig. 6.19c). By avoiding the use of light irradiation, this work opens a new therapy modality to tumor and pathogen infections.

A fluorescently detectable photosensitizer is beneficial for aiding in defining and adjusting parameters during PDT treatment. All these aforementioned works verified that CPEs are superior materials for PDT applications, which provided new opportunities to design CPEs based multimodal platform for both bioimaging and therapy applications in the future. Utilizing antibody or targeting protein conjugation, small targeting ligand conjugation, improved the targeting of PDT agents will be developed.

# References

1. Hamblin MR, Hasan T (2004) Photodynamic therapy: a new antimicrobial approach to infectious disease? Photochem Photobio Sci 3:436–450
2. Kobayashi H, Ogawa M, Alford R, Choyke PL, Urano Y (2009) New strategies for fluorescent probe design in medical diagnostic imaging. Chem Rev 110:2620–2640
3. Li P, Poon YF, Li W, Zhu H-Y, Yeap SH, Cao Y, Qi X, Zhou C, Lamrani M, Beuerman RW, Kang E-T, Mu Y, Li CM, Chang MW, Leong SSJ, Chan-Park MB (2011) A polycationic antimicrobial and biocompatible hydrogel with microbe membrane suctioning ability. Nat Mater 10:149–156
4. Lu L, Rininsland FH, Wittenburg SK, Achyuthan KE, McBranch DW, Whitten DG (2005) Biocidal activity of a light-absorbing fluorescent conjugated polyelectrolyte. Langmuir 21:10154–10159
5. Chemburu S, Corbitt TS, Ista LK, Ji E, Fulghum J, Lopez GP, Ogawa K, Schanze KS, Whitten DG (2008) Light-induced biocidal action of conjugated polyelectrolytes supported on colloids. Langmuir 24:11053–11062
6. Corbitt TS, Ding L, Ji E, Ista LK, Ogawa K, Lopez GP, Schanze KS, Whitten DG (2009) Light and dark biocidal activity of cationic poly(arylene ethynylene) conjugated polyelectrolytes. Photochem Photobio Sci 8:998–1005
7. Corbitt TS, Sommer JR, Chemburu S, Ogawa K, Ista LK, Lopez GP, Whitten DG, Schanze KS (2008) Conjugated polyelectrolyte capsules: light-activated antimicrobial micro "roach motels". ACS Appl Mater Interfaces 1:48–52
8. Zhou Z, Corbitt TS, Parthasarathy A, Tang Y, Ista LK, Schanze KS, Whitten DG (2010) "End-only" functionalized oligo(phenylene ethynylene)s: synthesis, photophysical and biocidal activity. J Phys Chem Lett 1:3207–3212
9. Tang Y, Corbitt TS, Parthasarathy A, Zhou Z, Schanze KS, Whitten DG (2011) Light-induced antibacterial activity of symmetrical and asymmetrical oligophenylene ethynylenes. Langmuir 27:4956–4962
10. Wang Y, Jett SD, Crum J, Schanze KS, Chi EY, Whitten DG (2012) Understanding the dark and light-enhanced bactericidal action of cationic conjugated polyelectrolytes and oligomers. Langmuir 29:781–792
11. Ding L, Chi EY, Chemburu S, Ji E, Schanze KS, Lopez GP, Whitten DG (2009) Insight into the mechanism of antimicrobial poly(phenylene ethynylene) polyelectrolytes:interactions with phosphatidylglycerol lipid membranes. Langmuir 25:13742–13751

12. Ding L, Chi EY, Schanze KS, Lopez GP, Whitten DG (2009) Insight into the mechanism of antimicrobial conjugated polyelectrolytes: lipid headgroup charge and membrane fluidity effects. Langmuir 26:5544–5550

13. Wang Y, Tang Y, Zhou Z, Ji E, Lopez GP, Chi EY, Schanze KS, Whitten DG (2010) Membrane perturbation activity of cationic phenylene ethynylene oligomers and polymers: selectivity against model bacterial and mammalian membranes. Langmuir 26:12509–12514

14. Wang Y, Chi EY, Natvig DO, Schanze KS, Whitten DG (2013) Antimicrobial activity of cationic conjugated polyelectrolytes and oligomers against saccharomyces cerevisiae vegetative cells and ascospores. ACS Appl Mater Interfaces 5:4555–4561

15. Zhu C, Yang Q, Liu L, Lv F, Li S, Yang G, Wang S (2011) Multifunctional cationic poly(p-phenylene vinylene) polyelectrolytes for selective recognition, imaging, and killing of bacteria over mammalian cells. Adv Mater 23:4805–4810

16. Wu C, Xu Q-H (2009) Enhanced one- and two-photon excitation emission of a porphyrin photosensitizes by fret from a conjugated polyelectrolyte. Macromol Rapid Commun 30:504–508

17. Zhang R-R, Li L, Tong L-L, Tang B (2013) Enhanced luminescence of photosensitizer-based mesoporous silica nanocomposites via energy transfer from conjugated polymer. Nanotechnology 24:015604–015604

18. Xing C, Xu Q, Tang H, Liu L, Wang S (2009) Conjugated polymer/porphyrin complexes for efficient energy transfer and improving light-activated antibacterial activity. J Am Chem Soc 131:13117–13124

19. Xing C, Yang G, Liu L, Yang Q, Lv F, Wang S (2012) Conjugated polymers for light-activated antifungal activity. Small 8:525–529

20. Liu L, Duan X, Liu H, Wang S, Li Y (2008) Microorganism-based assemblies of luminescent conjugated polyelectrolytes. Chem Commun 45:5999–6001

21. Liu L, Yu M, Duan X, Wang S (2010) Conjugated polymers as multifunctional biomedical platforms: Anticancer activity and apoptosis imaging. J Mater Chem 20:6942–6947

22. Xing C, Liu L, Tang H, Feng X, Yang Q, Wang S, Bazan GC (2011) Design guidelines for conjugated polymers with light-activated anticancer activity. Adv Funct Mater 21:4058–4067

23. Wang B, Yuan H, Zhu C, Yang Q, Lv F, Liu L, Wang S (2012) Polymer-drug conjugates for intracellar molecule-targeted photoinduced inactivation of protein and growth inhibition of cancer cells. Sci Rep 2:766

24. Salami S, Karami-Tehrani F (2003) Biochemical studies of apoptosis induced by tamoxifen in estrogen receptor positive and negative breast cancer cell lines. Clin Biochem 36:247–253

25. Rickert EL, Trebley JP, Peterson AC, Morrell MM, Weatherman RV (2007) Synthesis and characterization of bioactive tamoxifen-conjugated polymers. Biomacromolecules 8:3608–3612

26. Rivera-Guevara C, Perez-Alvarez V, Garcia-Becerra R, Ordaz-Rosado D, Sonia Morales-Rios M, Hernandez-Gallegos E, Cooney AJ, Elena Bravo-Gomez M, Larrea F, Camacho J (2010) Genomic action of permanently charged tamoxifen derivatives via estrogen receptor-alpha. Biorg Med Chem 18:5593–5601

27. Zhu C, Yang Q, Lv F, Liu L, Wang S (2013) Conjugated polymer-coated bacteria for multimodal intracellular and extracellular anticancer activity. Adv Mater 25:1203–1208

28. Yuan H, Chong H, Wang B, Zhu C, Liu L, Yang Q, Lv F, Wang S (2012) Chemical molecule-induced light-activated system for anticancer and antifungal activities. J Am Chem Soc 134:13184–13187

# Chapter 7
# Concluding Remarks

The integration of CPEs with biomedical research has already shown profound effects on the diagnosis for the early detection of diseases-related biomarkers, cell imaging, drug/gene delivery, and diseases treatment, which leads to a multidisciplinary scientific field in a context of chemistry, polymer science, materials, biology, and medicine with an integrated perspective into both basic research and application issues. All the CPEs presented in this Brief demonstrate that specific biomedical application can be achieved through delicate molecular design, which indicate that novel polymers modified with molecular fragments will be designed and synthesized with specific capability. To address the ever-changing and challenging needs of diagnosis, molecular imaging, drug delivery, and cancer therapy research, exploring novel CPEs with new features and capabilities are major issues in the ongoing work. The state-of-art synthetic methods used in this Brief provide the feasibility and flexibility to modify novel CPEs at the level of the conjugated repeat unit and pendant groups for improvement of the intrinsic CPE properties to enhance the detection sensitivity. Moreover, novel CPEs may be developed from borrowing better-known and newly discovered molecular targets of genomics and proteomics, which results in enlarging the species of targets. Benefiting from new important features, as well as inherent amplification effects, CPEs will provide a sensitive and convenient optical platform for highly specific bio/chemical sensing applications, and further for the early detection of diseases in complex biological fluids. Specific attention will be given to design CPEs-based multimodal platform through incorporating various reagents of specific functionalities, which holds great promises for various biomedical applications in the near future. It is believed that with the progress of in-depth research on CPEs-based biomedical application, more CPEs materials will be utilized to improve health care and advance biomedical research.

S. Wang and F. Lv, *Functionalized Conjugated Polyelectrolytes*, SpringerBriefs in Molecular Science, DOI: 10.1007/978-3-642-40540-2_7, © The Author(s) 2013